Alexander Karmazin

Time-efficient Simulation of Surface-excited Guided Lamb Wave Propagation in Composites

Karlsruher Institut für Technologie
Schriftenreihe des Instituts für Technische Mechanik

Band 18

Time-efficient Simulation of Surface-excited Guided Lamb Wave Propagation in Composites

by
Alexander Karmazin

Dissertation, Karlsruher Institut für Technologie (KIT)
Fakultät für Maschinenbau
Tag der mündlichen Prüfung: 19. Oktober 2012

Impressum

Karlsruher Institut für Technologie (KIT)
KIT Scientific Publishing
Straße am Forum 2
D-76131 Karlsruhe
www.ksp.kit.edu

KIT – Universität des Landes Baden-Württemberg und
nationales Forschungszentrum in der Helmholtz-Gemeinschaft

KIT Scientific Publishing 2012
Print on Demand

ISSN 1614-3914
ISBN 978-3-86644-935-0

Time-efficient Simulation of Surface-excited Guided Lamb Wave Propagation in Composites

Zur Erlangung des akademischen Grades

Doktor der Ingenieurwissenschaften

der
Fakultät für Maschinenbau
Karlsruher Institut für Technologie (KIT),

genehmigte
Dissertation

von

Dipl.-Math. Alexander Karmazin
aus Krasnodar, Russland

Tag der mündlichen Prüfung: 19. Oktober 2012
Hauptreferent: Prof. Dr.-Ing. Wolfgang Seemann
Korreferent: Prof. Dr. Evgenia Kirillova

Vorwort

Die vorliegende Arbeit enstand während meiner Mitarbeit an der Hochschule Rhein-Main und am Institut für Technische Mechanik des Karlsruher Instituts für Technologie in den Jahren 2007 bis 2012. An dieser Stelle möchte ich mich aus tiefstem Herzen bei all denjenigen bedanken, die zur Fertigstellung dieser Arbeit im universitären und im privaten Umfeld beigetragen haben.

Herrn Prof. Dr.-Ing. Wolfgang Seemann danke ich besonders für seine stets vorhandene Diskussionsbereitschaft, für die großzügige Unterstützung meiner Arbeit und für die Übernahme des Hauptreferates. Ganz besonders möchte ich mich auch für die mir gewährte Freiheit und entgegengebrachte Vetrauen bedanken.

Frau Prof. Dr. Evgenia Kirillova von der Hochschule RheinMain danke ich für die Anregung zu dieser Arbeit und die Übernahme des Korreferates. Ganz herzlich möchte ich für ihre intensive Betreuung und vertrauensvolle Förderung der Arbeit sowie für die hervorragende langjährige Zusammenarbeit danken. Dem Vorsitzenden des Prüfungsausschusses, Herrn Prof. Dr.-Ing. Xu Cheng vom Fachgebiet Fusionstechnologie und Reaktortechnik am KIT gilt ebenfalls mein Dank.

Mein aufrichtiger Dank gilt Herrn Prof. Dr.-Ing. Jörg Wauer für die stets vorhandene Diskussionsbereitschaft und die kritische Durchsicht des Manuskripts. Dem Dr. Pavel Syromyatnikov von der Staatlichen Kuban-Universität im Russland verdanke ich eine Reihe wertvoller Hinweise zur Verbesserung der Dissertation. Des Weiteren danke ich Herrn Prof. Dr. Evgeny Glushkov und Frau Prof. Dr. Natalya Glushkova ebenfalls von der Staatlichen Kuban-Universität für Ihre fachliche Unterstützung bei der Einarbeitung ins Thema.

Ebenfalls danke ich Herrn Prof. Dr.-Ing. Carsten Proppe, Herrn Prof. Dr.-Ing. Alexander Fidlin und Herrn Prof. Dr.-Ing. Dr. h.c. Jens Wittenburg für die wissenschaftliche Begleitung und ihre Diskussionsbereitschaft.

Den Kollegen am Institut für Technische Mechanik danke ich herzlich für die angenehme Arbeitsatmosphäre, für die interessanten und abwechslungsreichen Gespräche und die Hilfe bei den unzähligen organisatorischen Fragen. Insbesondere danke ich Dr.-Ing. Hartmut Hetzler, Dr.-Ing. Toni Liong, Dr.-Ing. Nantawatana Weerayuth, Dr.-Ing. Aydin Boyaci, M.-Eng. Xiaoyu Zhang und Dipl.-Wirtsch.-Ing. Marius Köster für die vielen wissenschaftlichen aber auch nicht-wissenschaftlichen Gespräche während unserer gemeinsamen Zeit am ITM.

Meiner Familie und insbesondere meinen Eltern Natalya und Vladimir danke ich von tiefstem Herzen für ihre fortwährende Unterstützung und den Rückhalt, auf den ich mich stets verlassen konnte.

Unendlich dankbar bin auch meiner lieben Frau Anna, die mich stets mit viel Geduld unterstützt und somit einen großen Anteil zum Gelingen dieser Arbeit beigetragen hat, auch wenn dies oft mit einem persönlichen Verzicht verbunden war.

Karlsruhe, den 15. März 2013

Alexander Karmazin

Kurzfassung

Viele Bauteile weisen die Eigenschaften eines Wellenleiters auf. Die bei der Wellenausbreitung entstehenden elastischen Deformationen können für die Auswertung der Struktureigenschaften verwendet werden, unter anderem für die zerstörungsfreie Werkstoffprüfung und für die Zustandsüberwachung. In der vorliegenden Arbeit werden die Methoden für eine rechenzeiteffiziente Berechnung der Wellenausbreitung in plattenförmigen, mehrschichtigen Verbundwerkstoffen vorgestellt. Dazu wird auf die partiellen Differentialgleichungen der Elastizitätstheorie mit zugehörigen Anfangs- und Randbedingungen des Wellenausbreitungsproblems die Fourier-Transformation angewandt. Die im Wellenzahl-Frequenz-Bereich ermittelte Lösung wird in Form einer Superposition der Greenschen Matrix und des Vektors der äußeren Anregung dargestellt. Für die Berechnung der Greenschen Matrix des Problems wird ein stabiler und zeiteffizienter numerischer Algorithmus verwendet. Die in den Komponenten der Greenschen Matrix auftretenden Polstellen werden für die Berechnung der Wellenzahlen der Lamb-Wellen in der gesamten Platte verwendet. Für die Ermittlung der dispersiven Eigenschaften der Wellen wird ein stabiles Verfahren vorgestellt, das die Berechnung von Phasen- und Gruppengeschwindigkeiten in Abhängigkeit von der Ausbreitungs- und Beobachtungsrichtung sowie auch der Frequenz erlaubt. Die damit berechneten numerischen Ergebnisse werden durch den Vergleich mit in der Literatur bekannten Daten validiert. Anhand der numerischen Ergebnisse werden auch die Grenzen der Anwendbarkeit der klassischen Plattentheorien von Kirchhoff und von Mindlin für das vorliegende Problem untersucht.

Im Weiteren wird ein Algorithmus für die Berechnung der Rücktransformation entwickelt, wobei die Lösung des Problems in Form einer Residuenreihe in den bereits ermittelten Polstellen dargestellt wird. Durch den Vergleich von damit berechneten Ergebnissen mit den Ergebnissen aus einer direkten Evaluation von Wellenzahlintegralen wird nachgewiesen, dass die Residuenreihendarstellung schon für die Entfernungen von ca. drei Wellenlängen von der Anregungsquelle gute qualitative und quantitative Ergebnisse liefert. Dabei ist die in dieser Arbeit entwickelte Methode wesentlich schneller und lässt die Ausbreitung der Wellenmoden getrennt betrachten. Außerdem wird eine asymptotische Entwicklung für die Residuenreihendarstellung hergeleitet. Daraufhin wird diese asymptotische Darstellung mit Hilfe der Airy-Funktion für den Fall der Existenz von Kaustiken erweitert. Anhand von umfangreichen numerischen Simulationen wird gezeigt, dass in den Fällen, bei denen eine starke Fokussierung der Wellen in einigen Richtungen zu beobachten ist, die asymptotische Entwicklung nur für große (von über dreißig Wellenlängen) Entfernungen von der Anregungsquelle angewandt werden kann. In den meisten anderen Fällen stimmen die Berechnungsergebnisse aller dieser Methoden schon für die Entfernungen von ca. fünf Wellenlängen von der Belastungsquelle gut überein. Darüber hinaus ist eine gute Übereinstimmung der mittels Residuenreihendarstellung berechneten Ergebnisse im Vergleich zu Ergebnissen der FEM-Simulationen sowie mit den experimentellen Daten zu konstatieren. Im Weiteren werden anhand zahlreicher Beispiele die in der vorliegenden Arbeit hergeleiteten Methoden verwendet, um Wellenausbreitungsphänomene und Energieprozesse zu untersuchen.

Abstract

Many typical construction parts have the properties of waveguides. The elastic deformations due to the elastic wave propagation are of great interest for nondestructive testing and structural health monitoring. In this thesis the methods of time-efficient simulation of surface-excited wave propagation in plate-like multilayered composites are presented. For obtaining the solution of wave propagation problem the Fourier transform is applied to the equations and boundary conditions of the corresponding mathematical model. In wavenumber-frequency domain the solution is derived as a product of the Green's matrix of the problem and a surface load vector. For the computation of Green's matrix representation in transformed domain a stable time-efficient algorithm is used. The poles of the components of Green's matrix are used for the computation of wavenumbers of Lamb waves in the whole composite plate in dependence on excitation frequency and observation direction. Then, the algorithm of computation of dispersive properties is validated by comparison of the group and phase velocities of Lamb waves excited and observed in the plate obtained numerically with data found in references. Furthermore, the limits of application of classical laminated plate and Mindlin laminated plate theories for wave propagation problems are studied using the obtained numerical results.

The next step is the development of the algorithm for the computation of the inverse Fourier transform. Here the solution of the problem is represented by the sum of residues corresponding to the poles calculated at the previous step. The displacements obtained applying this approach are compared with displacements obtained by the evaluation of two-dimensional wavenumber integral applying adaptive quadratures. The results let to conclude that the residue theorem-based approach gives good qualitative and quantitative coincidence of results already at distances of about three wavelengths from the excitation source. However, the residue-theorem based representation is essentially faster and allows to analyze the propagation of different wave modes separately. Moreover, the asymptotic expansion for the solution of the problem in far-field is developed. Then, this asymptotic expansion is expanded to the case of directions near caustics.

Based on numerical simulations it is shown that in cases of strong focussing of Lamb waves in some propagation directions the asymptotic expansion can be applied only for large (more than thirty wavelengths) distances from the excitation source. In most of other cases the results obtained applying all of these approaches are well coinciding already for distances about five wavelengths from the excitation source. Furthermore, a good coincidence of results obtained numerically applying the residue-theorem approach with results of FEM simulations and experimental data is observed. In addition the methods of computation of solution of wave propagation problem developed in this thesis are applied for studying the wave and energy propagation phenomena in composites.

Contents

1 Introduction

1.1 Background and motivation

In recent years composites become to be widely used in civil, mechanical, and aerospace engineering due to their high strength and lightness in comparison with metals. The use of composite materials allows to reduce the weight upto 30%. Moreover, composites have higher corrosion stability. Among others, composites are increasily used in aviation to reduce the mass of the airplanes and space ships. For example, the composite materials comprise more than 20% of the airframe of modern Airbus A380 [86]. On the other side, composites are complicated and more expensive in manufacturing. Another disadvantage consists of the sensitivity of composites to impact actions, i.e. small damages in the form of cracks or delaminations are practically unavoidable in manufacturing or in operation of the constructions made of composites. These small damages could potentially result in destruction of the construction.

Since small damages are not obvious and often cannot be detected immediately after the manufacturing of the component of the structure, the construction needs to be monitored in-situ or in relatively short intervals during its operation, when the construction is in the maintenance service. In the field of non-destructive testing (NDT) [18, 21, 22] and structural health monitoring (SHM) [14, 33, 132], there has been a growing interest in the recent decade in developing computer-aided systems for the detection of mechanical defects and the forecasting of destruction [133]. It is hoped that SHM systems will be able to regularly scan high-duty structural components and issue warnings concerning the formation of defects as well as provide an estimate of the remaining useful life. SHM systems can inform the user of the status of structure in real time and provide an estimate of the remaining useful life of the structure [122]. Moreover, the use of SHM systems can increase safety and can allow to change the maintenance procedure for aircraft from schedule driven to condition based, can reduce the fuel costs and the costs of maintenance significantly, and decrease the time required for the structure to be off-line [122].

In NDT following methods are widely used: eletrical, magnetic and electromagnetic (eddy current); sonic (ultrasonic); mechanical methods (vibration), optical (holography) and imaging methods (x-ray computer tomography), which all are well reviewed in [83]. However, not all of these methods can be performed while the structure is in service. The monitoring of the structure over time needs a network of sensors that are permanently attached onto the construction parts.

Since in many cases machine parts and structural components can be considered as waveguides, one of the promising approaches for SHM consists of applying the elastic waves propagating in the structures since the waves excited or reflected by damage provide significant information on the nature and properties of the defect. An improved inspection potential of guided waves over other ultrasonic methods is due to their sensitivity to different type of flaws, propagation over long distances and capability to follow curvature and reach hidden and/or buried part. The potential use of guided waves for monitoring metallic aircraft structures was investigated by Alleyne [4]. The level of understanding which has been reached in an application of elastic waves for NDT and SHM is documented in review works [21] and [110, 133], respectively. The references to numerous publications are provided.

The first NDT and SHM approach based on the use of elastic waves consists of the sensing and analysing the waves, excited by cracks during the initiation of the crack growth. This approach is passive and is named *acoustic emission* or *strain/loads monitoring*. The energy needs are low, however due to the low amplitudes of measuring responses a high density network of sensors is required [33]. The second approach uses an active scheme, where the nondestructive elastic waves are excited by external sources in a repeatable manner and the measured responses are used for the (quantitative) identification of the damages.

If the measurements are done at the same space point (if possible) where the waves were originally excited, such an approach is named *pulse-echo* technique. However, in this case the complicated process of wave reflection at structure boundaries needs to be known. Moreover, in large structures the amplitude of the waves reflected from boundaries is due to damping too low. Another way is to use *pitch-catch* technique, where the displacements are measured by another wafer than that used for wave excitation. However, clear understanding of quantitative connections between the waves and their sources is essential for the development of algorithms to detect defects. Moreover, the information about the structure in its undamaged state is required since by relating the measured response with one in the same but undamaged structure allows to considerably improve the precision of quantitative damage detection. This is shown, for example in case of the application of an approach based on damage influence maps in [70]. A clear understanding of quantitative connections between the waves and their sources would considerably ease the defect-feature recognition and would therefore enhance the outcome of the experiments and help to develop the necessary skill and experience in pattern recognition [124].

The first study of elastic wave propagation was done by Christoffel for the bulk waves in unbounded medium [24], followed by the study of surface waves in a half-space performed by Rayleigh [112]. However, many structures (e.g. aircraft panels) have the dimensions in one or two directions significantly larger than in other directions. Such structures can be considered as infinite layers (beams) or plates. Due to the finite dimensions of structural components the waves reflect from the structural boundaries, i.e. the waves guide through the structure. The first numerical results which relate to the characteristics of normal waves in a layer can be found in Lamb's works [71]. He was the first to obtain a dispersion equation linking frequencies and wave numbers. Hence, the waves in layer-like and plate-like structures are usually called (guided) Lamb waves.

The development of computer technology intensified the investigation of the properties of the propagation of elastic waves. Wave propagation in multilayered media with an arbitrary number of flat layers was derived for a plain strain ($2D$) problem by Thomson [137] and corrected by Haskell [46]. They connected the displacements and the stresses at the bottom of the layer with those at the top of the layer through a transfer matrix. An alternative formulation was used by Knopoff [65], where he connected the displacements and stresses at all layer interfaces including top and bottom boundaries by a global matrix. The first, who experimentally produced and measured Lamb waves was Worlton [152]. The first numerical calculations of real as well as complex branches of dispersion curves of Lamb modes in a traction-free plate were firstly published in Mindlin's [90] and Onoe's [101] works. Subsequent publications were dedicated to the study of the phenomenology of elastic waves and their propagation.

Not only the physical acoustics of Rayleigh and Lamb waves, but also their application for NDT was firstly studied by Viktorov [143]; however, only for isotropic elastic media. Theoretical principles of wave propagation in isotropic, anisotropic and layered materials, as well as of wave excitation with standard ultrasonic transducers for nondestructive evaluation, are described in works of Achenbach [2], Graff [45], Miklowitz [89], Auld [8], Rokhlin [115], Babeshko and Glushkov [10, 35], and Rose [116]. However, in these works the waveguides are modelled in two dimensions (plane strain problem), i.e. it requires that waves are excited by sources whose distribution is infinitely expanded in the direction perpendicular to the cross-section. To accurately model a finite-source induced wave propagation in anisotropic composite plates, a $3D$ formulation is required [141]. Nayfeh [94, 95] extended the Thomson-Haskell formulation to the case of $3D$-model of anisotropic material and composites of anisotropic layers, where the layers can have up to as low as monoclinic symmetry. The numerical instability of Thomson-Haskell method was resolved in works [63, 114, 148] by introducing the layer stiffness matrix and by using an efficient recursive algorithm to calculate the global stiffness matrix for an arbitrary anisotropic layered structure. In some works the Mindlin laminated plate theory [76, 136] or higher-order plate theories [150] are applied for modelling of free elastic waves in composites. Up to date, there are many methods developed for the computation of dispersion properties of laminated composites [79, 82, 149]. Analysis of dispersion properties of Lamb waves in anisotropic composites shows that additionally to the frequency dispersion the angular dispersion of waves should be taken into account. Moreover, some directions are privileged for the transport of energy of the guided waves, i.e. the waves are focussed [20, 40, 117].

Taking into account the dispersion properties of waves the problem of simulation of wave propagation excited by surface sources needs to be considered. Traditionally as wave actuation sources ultrasonic transducers are used for NDE purposes. However, due to their large geometrical dimensions, high weight and considerable power needed for actuation, for in-situ monitoring of structures low cost surface-coupled [33] or embedded piezoelectric actuators [70] are preferred. Due to the piezoelectric effect the wafers can be used not only as actuators but also as sensors. One of the most fundamental issues required for the effective use of piezoelectric actuators is the quantitative evaluation of the resulting elastic wave propagation by considering the coupled piezo-elastodynamic behavior between the actuator and the host medium as it is noted in [54]. However, under certain conditions the simplified models for surface sources can be applied [33]. More detailed information to the modelling of actuators is provided in a review work [54] or below in this thesis (section 2.4.3). Note that the problem of forced wave propagation is resolved applying the semi-analytical integral approach only for 2D-model for isotropic and anisotropic laminates [35, 37, 78] or 3D-model for isotropic laminates only [33, 108]. Inversion of expressions obtained in a 3D formulation in the transformed domain requires the computation of a double integral over wave numbers and a one-dimensional integral over frequencies, where the most computational effort is caused by double integral over wave numbers. The methods of the computation of this integral for the surface-excited wave propagation in anisotropic laminates are still time-inefficient [58, 84] or use an asymptotic representation of the wave field, which can be applied only at large distances to the excitation source [40, 43, 107, 141]. A more detailed review of these approaches is given in chapter 5 of this thesis.

Apart from the integral approach, the wave propagation problem can be resolved by applying direct numerical methods: conventional FEM [97, 98, 121], spectral FEM [30, 44, 70, 104], strip element method (SEM) [79, 125] and finite difference technique (FD) [126, 131]. Alleyne and Cawley [5] and Alleyne, et al [6] studied the propagation and scattering of Lamb waves in plates for nondestructive evaluation. In [103] is suggested the use of absorbing boundaries while employing the FEM and finite difference methods. A good review of all these direct numerical approaches is given in [72]. The application of direct numerical methods for the modelling of constructions from composites, is the most universal approach, as these allow to obtain an approximate solution for objects of any form. However, they are also the most expensive with regard to computational resources. The increase of the number of elements is unavoidable in regions of rapid changes of the solutions or characteristics of the medium (angular points, interfaces between contrast layers etc.) and especially in the case of high frequencies.

In some papers the FEM is used only for that part of a construction which has a complex form and comparable sizes in all directions. For the part of the construction which is a typical waveguide, i.e. for which the sizes in some directions considerably exceed those in other directions, the solution is constructed as a sum of propagating waves using the mode expansion technique (NME) [8]. Cho and Rose [23] combined the boundary element method and the normal-mode expansion method to study the edge reflection of Lamb waves and mode conversion with thickness variation. In [92] a combined FEM-NME approach is used for modelling the piezo-excited (plane strain problem) wave propagation in a composite plate. To use the NME technique for modelling the interaction of elastic waves with the piezoelectric actuator, Galan and Abascal [31] evaluated the Lamb wave propagation characteristics in sandwich plates in terms of absorbing boundary conditions derived from a truncated normal mode expansion technique. A hybrid method for layered structures named semianalytical finite element method (SAFE) is also suggested in [50] and used in [49, 50] for the analysis of wave processes in rails.

1.2 Research goal, scopes and objectives

The **goal** of **this research** is to understand, model, and predict the Lamb wave fields in laminated composite plates excited by surface stresses.

The **scope** of **this research** is to develop time-efficient predictive models of surface-excited guided Lamb wave propagation in anisotropic plate-like structures, to validate these models on various numerical examples, to study the properties of Lamb waves under harmonic and transient excitations, and to investigate the use of different piezoelectric actuators for wave actuation.

In detail, the **objectives** of this research are defined as follows:

- To present the accurate modelling of guided Lamb waves propagating in anisotropic multilayered plate-like structures by integral approach.

- To derive Green's matrix of the problem in wavenumber-frequency domain and study the dispersion properties of the structures in case of modelling of composite plates applying the elasticity theory and laminate plate theories of zero (Classical Laminated Plate Theory) and first (Mindlin Laminated Plate Theory) orders.

- To develop time-efficient methods for the evaluation of two-dimensional wave-number integral, which allow the computation of displacements and strains in both near- and far-fields to the excitation source with high accuracy.

- To estimate the computational error and study the limitations on the application of these methods.

- To demonstrate the application of the methods developed in this thesis for the prediction of harmonic and transient wave responses in composite plates under the excitation by piezoelectric wafers, where for wafers the simple pin-force model is used. To validate these methods by comparing the results of their application with the results of FEM simulations and with experimental data.

- To investigate the anisotropy-induced focussing of guided waves by computing the peak-to-peak amplitude curves and the power flow corresponding to each Lamb wave mode.

- To analyse the influence of the type of piezoelectric actuator on the resulting wave fields as well as to suggest some parameters for tuning (optimal wave excitation) of Lamb waves in composite plates under consideration.

1.3 Outline of thesis

This thesis studies the surface-excited guided Lamb waves in laminated composites. The study is carried out in a systematic approach consisting of several steps, each of them describes different aspects of the problem. These steps are organized in chapters in the following way.

Chapter 2 reviews the general properties of laminated composites, the mechanical models based on the elasticity theory and laminated plate theories, some known properties of Lamb waves propagating in multi- and singlelayered structures as well as the mechanical models describing the excitation and sensing of elastic waves.

In chapter 3 the boundary value problem of forced wave propagation in plate-like structures is transformed into the frequency-wavenumber domain. Then, the typical wavenumber domain representations of surface loads corresponding to pin-force models of different piezoelectric wafers are given. Finally, Green's matrix of the problem in frequency-wavenumber domain is constructed by applying a numerically stable algorithm, which does not contain any growing exponents. Owing to the real singularities, presented in Green's matrix components if no damping is considered, the unique solution of the problem in a time-space domain is obtained by choosing the integration contour in wavenumber domain according to the principle of limiting absorption.

Chapter 4 presents the stable algorithm of computing the dispersion curves of Lamb waves in composite plates, derived by using Green's matrix representation of the problem in frequency-wavenumber domain. Subsequently, this algorithm is applied for the investigation of dispersion properties of guided waves in laminated plates - phase and group velocity curves of incident waves as well as phase and group velocities of wavefronts (of observed waves), which are different due to the angular dispersion of waves. The dispersion curves are compared with numerical and experimental results in literature and a good agreement between all these curves is shown.

Chapter 5 is addressed to the development of effective methods of the evaluation of two-dimensional wavenumber integrals. After reviewing the techniques currently used for such integrals three approaches for the calculation of displacement fields in composite plates at different distances to the source are developed: the direct integration using adaptive quadratures for the near-field, "far-field residue integration" for the middle- and far-fields to the source, and asymptotic expansion for the significantly large distances to the source. The application of the "far-field residue integration" to the isotropic laminate under axis-symmetric loading is provided. It allows to make conclusion about the high accuracy of this approach. Moreover, it is observed that the asymptotic expansion obtained in this chapter is well coinciding with the formulas obtained previously by other authors. Then, this asymptotic expansion is extended to the case of directions near to caustics of Lamb waves. Finally, all approaches derived in this chapter are extensively tested on different numerical examples by varying the frequency, the actuator parameters and the composite stacking sequences.

In chapter 6 the algorithm of the computation of transient responses in composite plates under the excitation by surface sources is presented. A good coincidence of computational results with results of FEM simulations applying commercial FE software ABAQUS is observed. Moreover, the computational results are found to be in good agreement with experimental data published in literature. Using the approaches developed in previous chapter, the anisotropy-induced Lamb wave focussing is analysed by studying the amplitudes and power flows corresponding to the different wave modes. Finally, anti-resonance frequencies are studied. They appear in the case of out-of-phase excitation of wave modes by opposite boundaries of the actuator and are well-known for isotropic laminates [33]. However, due to the anisotropy of the plates, they occur not in all directions simultaneously. It results in the wave fields, where the central frequencies of waves observed in different directions are not equal.

Lastly, concluding remarks and an outlook on future studies are presented in the final chapter.

2 Modelling of free and forced wave motion in plate-like laminated composites

The objective of this chapter is to present the fundamentals of the modelling of excitation of Lamb waves in anisotropic laminated composites and their propagation within the composites. This basic information is needed hereafter to understand the methods and results provided in this work.

2.1 Summary of common properties of laminated composites

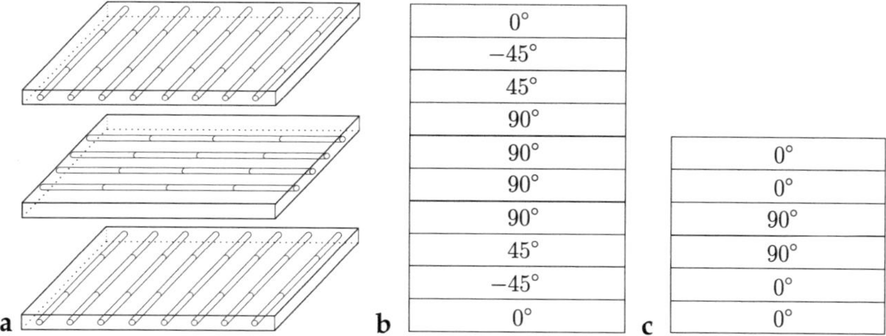

Figure 2.1: (a) Schematical representation of unidirectional layers characterized by fiber orientation. The composite layup of $[0/-45/45/90_2]_s$ (b) and of $[0_2/90]_s$ (c)

Laminated composite materials consist of layers of at least two different materials that are bounded together. Lamination is used to combine the best aspects of the constituent layers and bonding material in order to achieve a more useful material [56] (strength, stiffness, low weight, corrosion resistance, wear resistance, attractiveness, thermal insulation, acoustical insulation, etc.). Each layer in the laminated composite is a lamina, which is usually represented by an arrangement of unidirectional or woven fibers in an epoxy matrix. Typical materials used for fibers include carbon, glass and silicon carbide. The fibers are oriented in different directions to give different strengths and stiffnesses of the laminate in various directions (Figure 2.1a), i.e. to achieve specific design requirements on the structural element. If layers of different materials are used, such composites are called "hybrid". The most frequently used laminated composites are multilayered carbon-fiber-reinforced polymers (CFRP), glass fiber-reinforced plastics (GFRP) and ceramic matrix composites (CMCs). Other laminated composites widely used in industry are sandwich composites, i.e. hybrid laminates fabricated by attaching two thin but stiff layers (skins) to a lightweight but thick honeycomb or foam core. As skin materials not only CFRP and GFRP but also sheet metals are used.

2.1.1 Stacking sequence notation

The layers (or laminae) of the laminated composite can have different thicknesses, structure and mechanical properties [15]. For description of the properties of laminates composed of unidirectional plies of different mechanical properties and thicknesses the special stacking-sequence notations are used. For example, notation $[0/-45/+45/90_2]_s$ means that the laminate is symmetric with respect to mid-plane and the fibers in plies above the mid-plane are oriented in directions 0, -45, 45 and 90 with respect to one of the axes (usually x-axis of global coordinates) (see Figure 2.1b). The last layer above the mid-plane $[90]_2$ consists of two plies, each of them has the same thickness as the other three layers 0, -45 and 45. The stacking-sequence notation for this laminate can be also rewritten in an extended form as $[0/-45/+45/90/90]_s = [0/-45/45/90/90/90/90/45/-45/0]$. If the sequences of layers are repeated, they are grouped and indicated by a subscript, corresponding to the number of sequence repetitions [56], e.g. $[45/-45]_3 = [45/-45/45/-45/45/-45]$. In case of hybrid laminates the alphabetic characters are used in front of the fiber orientation angle to designate the material used for ply fabrication, e.g. $[C90/G0]_s$ means a symmetric composite laminate, where the outside layers are made of material "C" (carbon-epoxy) and the inside layers are made of material "G" (graphite-epoxy).

2.1.2 Typical composite layups

Laminated composites are inhomogeneous and anisotropic structures. The laminates are characterized by the elastic stiffnesses of the layers and by the geometry of the layup. With respect to both geometry and material properties, the laminates are devided in three groups: *symmetric, antisymmetric* and *unsymmetric* (also called *non-symmetric*) composites. Laminated composites without symmetry or antisymmetry of properties with respect to mid-plane are unsymmetric and are the most general class of laminates. However, symmetric composites due to the lack of the coupling between the extensional and bending motion are frequently used in various applications [56].

Examples of simple symmetric laminate with symmetric properties with respect to the middle plane are given by $[0/90/0] = [0/0/90/90/0/0] = [0_2/90]_s$ (Figure 2.1c) or by the single-layered composites. In cases, when the bending-extension coupling is of great importance for the practical application (e.g. airplane wings twisting under bending), antisymmetric composites are used. Antisymmetry of the laminate means that the sequence of layers below the middle surface of laminate is represented by a mirror image of the stacking sequence above the middle surface with signs of ply angles reversed. Antisymmetric laminates have an even number of layers with equal thicknesses, e.g. $[45/-45/45/-45]$. Below are briefly addressed some typical composite layups used in manufacturing of symmetric and antisymmetric laminated composites [56]:

- *unidirectional* composites consist of several layers with identical properties and under conditions of a continuity of stresses and displacements on the layer interfaces and hence they can be considered as single-layered composites;

- (regular) *cross-ply* composites are made of laminae of the same thickness and material properties, but have their major principal directions alternating at $0°$ and $90°$ to the laminate axes, for example $0/90/0$ (Figure 2.1c) (regular because of the same thicknesses and cross-ply because of the $90°$ angle between the fibers in adjacent layers);

- (regular) *angle-ply* composites consist of layers of alternating fiber orientations α and $-\alpha$ to axial direction of the laminate, e.g. $[\alpha/-\alpha/\alpha]$ or $[30/-30/30/-30]_s$;

- *quasi-isotropic* composites have isotropic extensional stiffnesses (the same in all directions in the plane of the laminate). This means that the laminate, in some sense, appears isotropic, but is not actually isotropic in all senses, because under transverse and interlaminar shear loading its behaviour differs from that of an isotropic layer [15]. Some simple examples of quasi-isotropic laminates are $[-60/0/60]$ and $[0/-45/45/90]$.

Also note that *hybrid* laminates are not unidirectional, angle-ply or quasi-isotropic, because the laminae are of different materials.

2.1.3 Generalized Hooke's law

The individual layers in the laminate can be generally considered as having common anisotropic (or triclinic) properties. The stresses σ_{ij} in generally anisotropic material in case of negligible damping are related to strains ε_{ij} by a generalized Hooke's law[12]:

$$\sigma_{ij} = C_{ijkl}\varepsilon_{kl}, \quad i,j,k,l = 1,2,3. \tag{2.1}$$

---

[1]In this work all vectors are denoted using bold font and their modulus or components are expressed in normal font.

[2]If not otherwise stated, the Einstein summation convention of summing on repeated indices is used throughout the work.

Taking into account the symmetries $C_{ijkl} = C_{klij} = C_{jikl} = C_{ijlk}$, $\sigma_{ij} = \sigma_{ji}$ and $\varepsilon_{ij} = \varepsilon_{ji}$ of tensors in (2.1), matrix form (also called contracted (Voigt) notation) of generalized Hooke's law can be obtained. It describes the six independent components of stress or strain, which use a pair of indices ranging from 1 to 3, by a single index which ranges from 1 to 6. The indices are corresponded as follows: $11 \leftrightarrow 1$, $22 \leftrightarrow 2$, $33 \leftrightarrow 3$, $23 \leftrightarrow 4$, $31 \leftrightarrow 5$, $12 \leftrightarrow 6$ [74]:

$$\sigma_i = C_{ij}\varepsilon_j, \quad i = 1,\ldots,6. \tag{2.2}$$

Explicit form of the stiffness matrix and of the stress and strain vectors in (2.2) is stated in Appendix A.1. Note that the stiffness matrix $\mathbf{C}$ of size 6×6 in (2.2) should not be confused with the stiffness tensor $\mathcal{C}$ of size $3 \times 3 \times 3 \times 3$ in (2.1). The stiffness matrix $\mathbf{C}$ is symmetric and has at most 21 independent stiffness components. Material symmetry allows to reduce the number of independent components. For example, in laminated composites the individual layers generally are orthotropic (with two orthogonal planes of material property symmetry, i.e. 9 independent stiffnesses) or transversely isotropic (properties are equal in one of the planes in all directions, i.e. 5 independent stiffnesses). For description of such materials as well as for isotropic structures instead of stiffness matrix alternatively the engineering constants are used. Details on the computation of the stiffness matrix using engineering constants are provided in Appendix A.4.

Both Equations (2.1), (2.2) are written in terms of some local coordinates x_1, x_2 and x_3 associated with the material principal axes. However, if the global coordinate system x_1', x_2' and x_3' does not coincide with principal coordinates of the stiffness tensor $\mathcal{C}$, the coordinates of the tensor in x_1', x_2', x_3' are obtained according to [74]

$$C'_{ijkm} = a_{pi}a_{qj}a_{rk}a_{ms}C_{pqrs}, \tag{2.3}$$

where C_{pqrs} are the coordinates of the stiffness tensor with respect to local coordinate system, and C'_{ijkm} are the coordinates of the stiffness tensor with respect to the global coordinate axes. This formula and corresponding transformation rule in terms of stiffness matrix is given in Appendix A.2. Usually the formula is needed to obtain the stiffness constants of the layer in some global coordinate system using its stiffnesses given in its local coordinate system. Thus, for example considering a unidirectional composite layer with fibres located at an angle β with respect to 1-axis of global coordinate system, corresponding global stiffness constants are obtained using the rotation matrix

$$\mathbf{a} = \begin{pmatrix} \cos\beta & -\sin\beta & 0 \\ \sin\beta & \cos\beta & 0 \\ 0 & 0 & 1 \end{pmatrix}. \tag{2.4}$$

Such a coordinate transformation is needed usually to obtain the stress-strain relations of the laminated composite in global coordinate directions, geometrically convenient to the solution of the problem.

2.2 Equations of motion

In this section the equations of motion in a laminated composite are given in terms of displacements using the 3D elasticity theory and approximate plate theories. The multilayered laminated structure considered in this section is represented schematically in Figure 2.2.1.

2.2.1 Elastodynamic equations of $3D$ elasticity theory

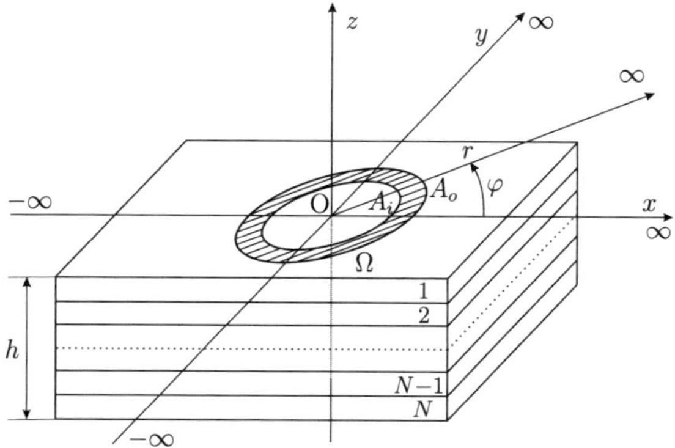

Figure 2.2: N-layered plate under an excitation by a ring-shaped source (inner radius A_i, outer radius A_o

In the case of absence of body forces the motions in the n-th layer, $n = 1, \ldots, N$ are expressed in terms of stress and displacement components with respect to a rectangular Cartesian coordinate system x, y and z[1]:

$$\frac{\partial \sigma_x^{(n)}}{\partial x} + \frac{\partial \tau_{xy}^{(n)}}{\partial y} + \frac{\partial \tau_{xz}^{(n)}}{\partial z} = \varrho^{(n)} \frac{\partial^2 u_x^{(n)}}{\partial t^2},$$

$$\frac{\partial \tau_{xy}^{(n)}}{\partial x} + \frac{\partial \sigma_y^{(n)}}{\partial y} + \frac{\partial \tau_{yz}^{(n)}}{\partial z} = \varrho^{(n)} \frac{\partial^2 u_y^{(n)}}{\partial t^2}, \tag{2.5}$$

$$\frac{\partial \tau_{xz}^{(n)}}{\partial x} + \frac{\partial \tau_{yz}^{(n)}}{\partial y} + \frac{\partial \sigma_z^{(n)}}{\partial z} = \varrho^{(n)} \frac{\partial^2 u_z^{(n)}}{\partial t^2},$$

where $\mathbf{u}^{(n)} = (u_x^{(n)}, u_y^{(n)}, u_z^{(n)})^T$ is the displacement vector of n-th layer, t is the time variable and $\varrho^{(n)}$ is the mass density of the n-th layer. These governing equations

---

[1]In the following two notations for the global coordinate system are assumed to be equivalent $x_1 = x$, $x_2 = y$ and $x_3 = z$. Also the use of indices x and 1, y and 2, z and 3 for denoting the components of vectors is equivalent.

of linear elasticity in the case of absence of body forces are also known as Cauchy's first equations of motion or *elastodynamic equations*. However, to obtain a displacement formulation of Equations (2.5), stresses are substituted by strains using Hooke's law (2.2). In turn, the strains ε_{ij} are given through displacements by

$$\varepsilon_{ij} = \frac{1}{2}\left(u_{j,i} + u_{i,j}\right). \tag{2.6}$$

This yields the displacement equations of motion for each layer[1] in matrix form

$$(\mathbf{A}^{(01)}\frac{\partial^2}{\partial x^2} + \mathbf{A}^{(02)}\frac{\partial^2}{\partial y^2} + \mathbf{A}^{(03)}\frac{\partial^2}{\partial x\partial y} + \mathbf{A}^{(04)})\mathbf{u} \tag{2.7}$$

$$+(\mathbf{A}^{(11)}\frac{\partial}{\partial x} + \mathbf{A}^{(12)}\frac{\partial}{\partial y})\mathbf{u}' + \mathbf{A}^{(2)}\mathbf{u}'' = 0,$$

where $\mathbf{u}' = \partial\mathbf{u}/\partial z$ and $\mathbf{u}'' = \partial^2\mathbf{u}/\partial z^2$ denote the first and second partial derivatives of the displacement vector $\mathbf{u} = (u_x, u_y, u_z)^T$ with respect to z. All matrices in (2.7) are of size 3×3 and are dependent on stiffness components C_{ij} of the layer, density ϱ of the layer. Further the matrix $\mathbf{A}^{(04)}$ includes an operator corresponding to the second order partial derivative with respect to t. More detailed form of these matrices is given in Appendix A.5.

2.2.2 Boundary and initial conditions for the elastodynamic problem

The elastodynamic problem formulation is completed by the boundary conditions. The structural elements needed in SHM have in-plane dimensions much larger than their thickness and therefore can be considered as plate-like structures of infinite horizontal dimensions. Thus, the laminated composite occupies the volume $-\infty \le x,y \le \infty$, $z_{N+1} \le z \le 0$, where z_1, z_{N+1} and z_n ($n = 2,\dots,N$) - are the z-coordinates of the upper, bottom and interface surfaces of the laminate respectively. The thickness of the laminated plate is denoted as h within the whole thesis. Conventionally, the origin of the global coordinate system is assumed to be at the upper surface of the laminate [10, 109, 127], i.e. $z_1 = 0$ and $z_{N+1} = -h$. Nevertheless, in some cases it is more convenient to place the origin at the middle surface of the laminate [15, 76], i.e. $z_1 = h/2$ and $z_{N+1} = -h/2$. In the following both coordinate systems are used, if needed, the choice of the coordinate system in a concrete case will be discussed.

If the laminate does not contain any delaminations or cracks, the stresses σ_{j3} and displacements u_j are continuous on the layer interfaces

$$\sigma_{j3}^{(n)} = \sigma_{j3}^{(n+1)}, \quad u_j^{(n)} = u_j^{(n+1)}, \quad z = z_{n+1}, \quad n = 1,\dots,N-1. \tag{2.8}$$

---

[1]Index of the layer number n is hereinafter omitted for simplicity.

This condition is assumed within the whole work. In the following the problem of free elastic wave motion in the laminate is considered for the plate with traction-free (or stress-free) upper and lower boundaries (i.e. the surrounding medium of the plate is the vacuum)

$$\left(\sigma_{13}^{(1)},\sigma_{23}^{(1)},\sigma_{33}^{(1)}\right)\Big|_{z=z_1} = 0, \tag{2.9}$$

$$\left(\sigma_{13}^{(N)},\sigma_{23}^{(N)},\sigma_{33}^{(N)}\right)\Big|_{z=z_{N+1}} = 0. \tag{2.10}$$

However, in some problems the lower boundary is clamped[1] and corresponding boundary conditions are

$$\left(u_1^{(N)},u_2^{(N)},u_3^{(N)}\right)\Big|_{z=z_{N+1}} = 0. \tag{2.11}$$

The main objective of this work is to investigate the propagation of surface excited waves in a laminated plate with traction-free lower boundary (2.10). The excitation force is assumed to be applied in the domain Ω at the upper surface ($z = z_1$), replaces[2] (2.9) and is described as given surface tractions $q_j(x,y)$ and time-dependent excitation pulse $V(t)$:

$$\sigma_{j3}^{(1)} = \begin{cases} V(t)q_j(x,y), & j = 1,2,3, \quad (x,y) \in \Omega, \\ 0, & j = 1,2,3, \quad (x,y) \notin \Omega. \end{cases} \tag{2.12}$$

Note that due to the linearity of the elastodynamic problem (2.5), (2.8), (2.10), (2.12) the action of several surface sources results in displacement field represented by the sum of corresponding displacement fields for each source.

Due to the infinite horizontal dimensions of the plate, it is assumed that the displacement vector $\mathbf{u}(\mathbf{x},t)$ tends to zero at infinity, i.e.

$$\mathbf{u}(\mathbf{x},t) \to 0, \quad \text{if} \quad x,y \to \infty. \tag{2.13}$$

Also zero displacements and velocities at time $t = 0$ are assumed:

$$\mathbf{u}(\mathbf{x},t)\big|_{t\leq 0} = \mathbf{0}, \quad \frac{\partial \mathbf{u}(\mathbf{x},t)}{\partial t}\bigg|_{t\leq 0} = \mathbf{0}. \tag{2.14}$$

---

[1] Note that in case of the fixed lower boundary the waves are not exactly Lamb waves studied in this work but waves of similar type [10].

[2] Also an excitation source acting at the bottom surface of composite can be taken into account in a similar way.

2.2.3 Mindlin Laminated Plate Theory

In order to simplificate a problem, instead of elastodynamic equations (2.5) and boundary conditions (2.10), (2.12) other simplified models for composite materials can be used, as it is stated in work [19]. In this section the approximate solutions of the Mindlin first-order shear deformation plate theory (also known as Mindlin Laminated Plate Theory (MLPT)) are used. The Mindlin plate theory assumes that under deformation the normal to the mid-surface of plate remains straight but not necessarily perpendicular to the mid-surface. The in-plane displacements in the plate vary linearly through the thickness and the out-of-plane displacement does not change through the thickness [15]:

$$\begin{aligned}
u(x,y,z,t) &= u_0(x,y,t) + z\psi_x(x,y,t), \\
v(x,y,z,t) &= v_0(x,y,t) + z\psi_y(x,y,t), \\
w(x,y,z,t) &= w_0(x,y,t),
\end{aligned}$$

(2.15)

where u_0, v_0 and w_0 represent the displacements of every point of the mid-plane, ψ_x and ψ_y correspond to the rotations of sections $x = $ const and $y = $ const respectively. In (2.15) the center of global coordinate system of the problem as opposed to elastodynamic problem (2.5) is chosen to be located at the mid-surface of the composite plate. By substituting relations (2.15) into strain-displacement relations given in (A.1),

$$\begin{aligned}
\varepsilon_x &= u_{0,x} + z\psi_{x,x}, \quad \varepsilon_y = v_{0,y} + z\psi_{y,y}, \quad \varepsilon_z = 0, \\
\gamma_{xy} &= u_{0,y} + v_{0,x} + z\left(\psi_{x,y} + \psi_{y,x}\right), \quad \gamma_{xz} = \psi_x + w_{0,x}, \quad \gamma_{yz} = \psi_y + w_{0,y}
\end{aligned}$$

(2.16)

is obtained[1]. According to these strain-displacement relations, the shear strain is constant across the thickness of the plate. But the shear stress is known to be parabolic and to correct the discrepancies between the actual displacement field (elastodynamic equations) and that of the plate theory, shear strains are corrected by introducing shear correction factors κ_1, κ_2. The corresponding shear strains therefore are replaced by

$$\gamma_{xz} = \kappa_1\left(\psi_x + w_{0,x}\right), \quad \gamma_{yz} = \kappa_2\left(\psi_y + w_{0,y}\right).$$

(2.17)

The shear correction factors κ_1, κ_2 are to be determined. They can be chosen from energy considerations or alternatively, they can be set by matching the cut-off frequency of the thickness-shear motion in the low frequency range. For an isotropic homogeneous plate they are obtained as $\kappa_1 = \kappa_2 = \sqrt{2/3}$ or by matching cut-off frequency $\kappa_1^2 = \kappa_2^2 = \pi^2/12$ [16]. In case of an orthotropic plate, shear correction factors are found to be $\kappa_1 = \kappa_2 = \sqrt{5/6}$. In case of a multilayered composite plate, they can be found, generally speaking, only numerically using formulas provided in [15].

Additionally, the Mindlin plate theory assumes the normal stresses negligible within the volume of plate, i.e. a plane stress problem ($\sigma_z = 0$) is considered. Instead of the

---

[1]Hereinafter, the subindex $,x$ denotes a partial derivative with respect to x variable.

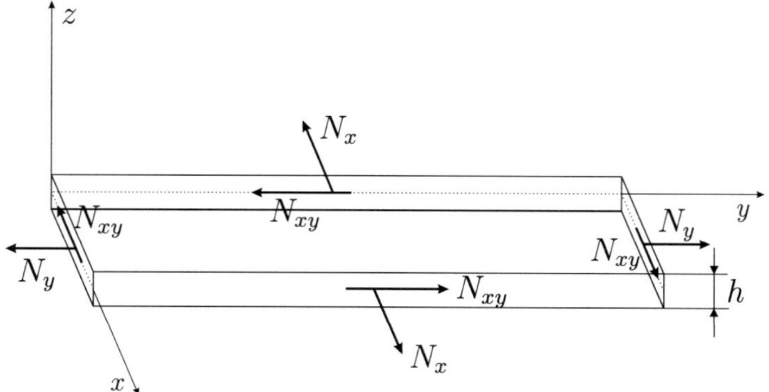

Figure 2.3: In-plane stress resultants applied to a laminate element [16]

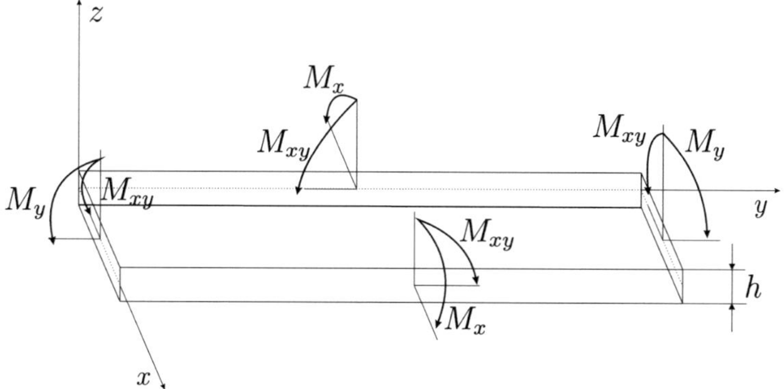

Figure 2.4: Moment resultants applied to a laminate element [16]

stresses in the elasticity problem (2.5) in the plate theory, stress and moment resultants are used. Stress resultants N_x, N_y and N_{xy} are given as

$$\left(N_{ij}\right) = \int_{-h/2}^{h/2} \sigma_{ij}\,\mathrm{d}z, \quad i,j = 1,2 = x,y, \tag{2.18}$$

the moment resultants M_x, M_y and M_{xy} as

$$\left(M_{ij}\right) = \int_{-h/2}^{h/2} z\sigma_{ij}\,\mathrm{d}z, \quad i,j = 1,2 = x,y, \tag{2.19}$$

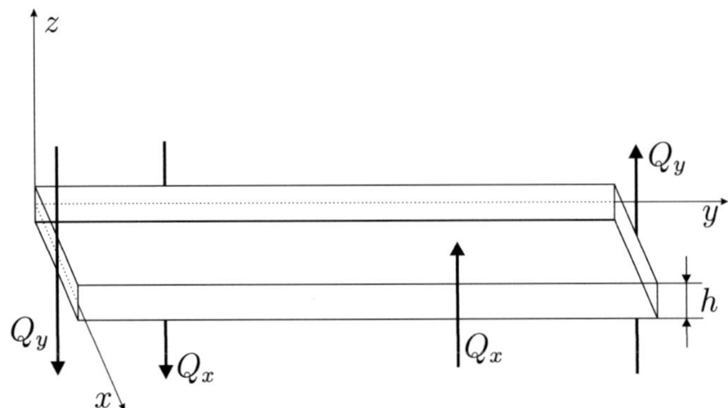

Figure 2.5: Shear resultants applied to a laminate element [16]

and the shear resultants Q_x and Q_y as

$$(Q_i) = \kappa_i \int_{-h/2}^{h/2} \sigma_{i3}\, dz, \quad i = 1, 2 = x, y.$$

(2.20)

The constitutive equations for a linear elastic multilayered plate are derived from the 3D elasticity theory taking into account corrections made in (2.17) and to be obtained in the following matrix form:

$$\begin{pmatrix} \mathbf{N} \\ \mathbf{M} \end{pmatrix} = \begin{pmatrix} \mathbf{A} & \mathbf{B} \\ \mathbf{B} & \mathbf{A} \end{pmatrix} \begin{pmatrix} \boldsymbol{\varepsilon}^{(0)} \\ \boldsymbol{\varepsilon}^{(1)} \end{pmatrix}, \quad \mathbf{Q} = \mathbf{H}\boldsymbol{\gamma}^{(0)},$$

(2.21)

where the stress $\mathbf{N}$, moment $\mathbf{M}$ and shear $\mathbf{Q}$ vectors are defined as

$$\mathbf{N} = \left(N_x, N_y, N_{xy}\right)^T, \quad \mathbf{M} = \left(M_x, M_y, M_{xy}\right)^T, \quad \mathbf{Q} = \left(Q_x, Q_y\right)^T,$$

(2.22)

and the vectors $\boldsymbol{\varepsilon}^{(0)}$, $\boldsymbol{\varepsilon}^{(1)}$ and $\boldsymbol{\gamma}^{(0)}$ as

$$\begin{aligned}
\boldsymbol{\varepsilon}^{(0)} &= \left(u_{0,x}, v_{0,y}, u_{0,y} + v_{0,x}\right)^T, \quad \boldsymbol{\varepsilon}^{(1)} = \left(\psi_{x,x}, \psi_{y,y}, \psi_{x,y} + \psi_{y,x}\right)^T, \\
\boldsymbol{\gamma}^{(0)} &= \left(\psi_x + w_{0,x}, \psi_y + w_{0,y}\right)^T,
\end{aligned}$$

(2.23)

and the submatrices $\mathbf{A}$, $\mathbf{B}$, $\mathbf{D}$ and $\mathbf{H}$ as

$$\mathbf{A} = \begin{pmatrix} A_{11} & A_{12} & A_{16} \\ A_{12} & A_{22} & A_{26} \\ A_{16} & A_{26} & A_{66} \end{pmatrix}, \quad \mathbf{B} = \begin{pmatrix} B_{11} & B_{12} & B_{16} \\ B_{12} & B_{22} & B_{26} \\ B_{16} & B_{26} & B_{66} \end{pmatrix},$$

(2.24)

$$\mathbf{D} = \begin{pmatrix} D_{11} & D_{12} & D_{16} \\ D_{12} & D_{22} & D_{26} \\ D_{16} & D_{26} & D_{66} \end{pmatrix}, \quad \mathbf{H} = \begin{pmatrix} \kappa_1^2 A_{55} & \kappa_1 \kappa_2 A_{45} \\ \kappa_1 \kappa_2 A_{45} & \kappa_2^2 A_{44} \end{pmatrix}.$$

The elements of submatrices $\mathbf{A}$, $\mathbf{B}$ and $\mathbf{D}$ are evaluated for $i, j = 1, \ldots, 6$ as follows

$$A_{ij} = \int\limits_{-h/2}^{h/2} \overline{P}_{ij}(z)\, dz, \quad B_{ij} = \int\limits_{-h/2}^{h/2} \overline{P}_{ij}(z)\, z\, dz, \quad D_{ij} = \int\limits_{-h/2}^{h/2} \overline{P}_{ij}(z)\, z^2\, dz, \tag{2.25}$$

where the $\overline{P}_{ij}$ are given by the reduced stiffness matrices $\mathbf{P}^k$ of each layer as

$$\overline{P}_{ij}(z) = P_{ij}^k, \text{ if } z_k \leq z \leq z_{k+1}, \tag{2.26}$$
$$P_{ij}^k = C_{ij}^k - C_{i3}^k C_{j3}^k / C_{33}^k, \quad i, j = 1, 2, 6.$$

The integration of the elastodynamic equations (2.5) through the thickness and the substitution of stress (2.18) and shear resultants (2.20) into obtained equations leads to the fundamental equations for in-plane and transverse shear resultants. The fundamental equations for the moments are obtained by multiplying the first two equations in (2.5) by z, integrating over the thickness and substituting the moment resultants (2.19) [15]. Thus, the following system is obtained[1]:

$$\begin{aligned}
\frac{\partial N_x}{\partial x} + \frac{\partial N_{xy}}{\partial y} &= I_0 \ddot{u}_0 + I_1 \ddot{\psi}_x + q_1(x, y) V(t), \\[2mm]
\frac{\partial N_{xy}}{\partial x} + \frac{\partial N_y}{\partial y} &= I_0 \ddot{v}_0 + I_1 \ddot{\psi}_y + q_2(x, y) V(t), \\[2mm]
\frac{\partial Q_x}{\partial x} + \frac{\partial Q_y}{\partial y} &= I_0 \ddot{w}_0 + q_3(x, y) V(t), \\[2mm]
\frac{\partial M_x}{\partial x} + \frac{\partial M_{xy}}{\partial y} - Q_x &= I_1 \ddot{u}_0 + I_2 \ddot{\psi}_x + \frac{h}{2} q_1(x, y) V(t), \\[2mm]
\frac{\partial M_{xy}}{\partial x} + \frac{\partial M_y}{\partial y} - Q_y &= I_1 \ddot{v}_0 + I_2 \ddot{\psi}_y + \frac{h}{2} q_2(x, y) V(t),
\end{aligned} \tag{2.27}$$

where

$$I_j = \int\limits_{-h/2}^{h/2} \varrho(z) z^j\, dz = \frac{1}{j+1} \sum_{k=1}^{N} \varrho_k (z_{k-1}^{j+1} - z_k^{j+1}), \quad j = 0, 1, 2, \tag{2.28}$$

and overdots denote partial derivatives with respect to time t. The term I_0 is the weight per unit area of the laminate. The last two equations in (2.27) introduce the coefficients I_1 and I_2, which are called the rotational inertia terms.

Equations (2.27) associated to the boundary conditions of the structure have five unknown functions $u_0(x, y, t)$, $v_0(x, y, t)$, $w_0(x, y, t)$, $\psi_x(x, y, t)$ and $\psi_y(x, y, t)$, which are the solution of the elasticity problem for Mindlin laminated plate. Corresponding

---

[1]Note that the boundary conditions (2.12), (2.10) are taken into account while integrating along the thickness.

Equations (2.27) can be formulated in an explicit matrix form with respect to unknown vector-function $\mathbf{u}_M = \left(u_0, v_0, w_0, \psi_x, \psi_y\right)^T$ as it is done in [76]

$$\mathbf{T}_M \mathbf{u}_M = \mathbf{f}_M, \tag{2.29}$$

where the matrix $\mathbf{T}_M$ contains partial derivatives with respect to x, y and t and material properties, given by components of the matrices $\mathbf{A}$, $\mathbf{B}$, $\mathbf{D}$, $\mathbf{H}$ and components of the vector $\mathbf{I}$ (2.24),(2.28). The explicit form of the matrix $\mathbf{T}_M$ is stated in Appendix 2.2.3. The load vector in (2.29) is given in terms of the surface load vector (2.12) as

$$\mathbf{f}_M = V(t) \left(q_1(x,y), q_2(x,y), q_3(x,y), \frac{h}{2} q_1(x,y), \frac{h}{2} q_2(x,y) \right)^T. \tag{2.30}$$

Note that the reduced stiffness matrix $\mathbf{B}$ represents the coupling of symmetric and antisymmetric Lamb wave modes. In case of a laminated composite plate symmetric with respect to mid-plane, the matrix $\mathbf{B} = \mathbf{0}$, rotational inertia term $I_1 = 0$, and Equation (2.29) is simplified and the in-plane and flexural motions of the laminated plate are uncoupled.

The first-order shear deformation plate theory formulated in this section gives an approximation of the equations of motion of $3D$ elasticity theory (2.5) and allows quick analysis of the Lamb wave propagation problem. However, as it is known already for isotropic solids, the MLPT is valid only at low frequency range. In order to get better results for higher frequencies, the MLPT can be extended to a higher-order plate theory as it is done in [113, 150].

2.2.4 Classical Laminate Plate Theory

At very low frequencies the shear deformations are negligible and instead of the Mindlin plate theory (2.29), the Classical Laminated Plate Theory (CLPT) can be used. The CLPT assumes that the deformation of the normal to the mid-plane is then a straight line normal to the deformed mid-plane [15] and neglects the effect of the transverse shear:

$$\gamma_{xz} = 0, \quad \gamma_{yz} = 0. \tag{2.31}$$

It implies from (2.16) that

$$\psi_x(x,y) = -\frac{\partial w_0}{\partial x}, \quad \psi_y(x,y) = -\frac{\partial w_0}{\partial y}. \tag{2.32}$$

The displacement field is then rewritten as

$$\begin{aligned}
u(x,y,z,t) &= u_0(x,y,t) - z\frac{\partial w_0}{\partial x}(x,y,t), \\
v(x,y,z,t) &= v_0(x,y,t) - z\frac{\partial w_0}{\partial y}(x,y,t), \\
w(x,y,z,t) &= w_0(x,y,t).
\end{aligned} \tag{2.33}$$

Strain components taking into account (2.31) are found to be

$$\varepsilon_x = u_{0,x} - zw_{0,xx}, \quad \varepsilon_y = v_{0,y} - zw_{0,yy}, \quad \varepsilon_z = 0,$$
$$\gamma_{xy} = u_{0,y} + v_{0,x} - 2zw_{0,xy}, \quad \gamma_{xz} = 0, \quad \gamma_{yz} = 0.$$

(2.34)

Substitution of relation (2.32) into equations of motion of Mindlin plate yields that shear stresses are absent: $\mathbf{Q} = \mathbf{0}$. Moreover, the number of independent unknown functions reduces to three and the system of equations (2.27) takes the form [15]

$$\frac{\partial N_x}{\partial x} + \frac{\partial N_{xy}}{\partial y} = I_0 \ddot{u}_0 - I_1 \ddot{w}_{0,x} + q_1(x,y)V(t),$$

$$\frac{\partial N_{xy}}{\partial x} + \frac{\partial N_y}{\partial y} = I_0 \ddot{v}_0 - I_1 \ddot{w}_{0,y} + q_2(x,y)V(t),$$

(2.35)

$$\frac{\partial^2 M_x}{\partial x^2} + 2\frac{\partial^2 M_{xy}}{\partial x \partial y} + \frac{\partial^2 M_y}{\partial y^2} = I_0 \ddot{w}_0 + I_1 \left(\ddot{u}_{0,x} + \ddot{v}_{0,y} \right)$$

$$- I_2 \left(\ddot{w}_{0,xx} + \ddot{w}_{0,yy} \right) + q_3(x,y)V(t).$$

Generally the rotational inertia terms can be neglected: $I_1 = I_2 = 0$. Rewriting the equations of motion in a matrix form similar to (2.29) gives

$$\mathbf{T}_C \mathbf{u}_C = \mathbf{q}.$$

(2.36)

The matrix $\mathbf{T}_C$ is given in an explicit form in Appendix A.7. The vector of unknown displacements is $\mathbf{u}_C = (u_0, v_0, w_0)$.

Note that for a symmetric plate $\mathbf{B} = \mathbf{0}$, the equations of motion (2.36) can be considerably simplified. Moreover, the in-plane behaviour (u_0, v_0) is decoupled from the flexural behaviour (w_0) and the corresponding equation of out-of-plane motion can be solved independently [15]. For example, in the case of an orthotropic plate of thickness h (unidirectional composite) with x-axis taken parallel to the fiber direction, the following equation for out-of-plane motion is obtained

$$D_{11}\frac{\partial^4 w_0}{\partial x^4} + 2(D_{12} + 2D_{66})\frac{\partial^4 w_0}{\partial x^2 \partial y^2} + D_{22}\frac{\partial^4 w_0}{\partial y^4} + I_0\frac{\partial^2 w_0}{\partial t^2} = q_3(x,y)V(t),$$

(2.37)

where the coefficients are expressed through the stiffnesses of the plate (2.24)

$$D_{11} = \frac{h^3}{12}\frac{C_{11}C_{33} - C_{13}^2}{C_{33}}, \quad D_{22} = \frac{h^3}{12}\frac{C_{22}C_{33} - C_{23}^2}{C_{33}},$$

(2.38)

$$D_{12} + 2D_{66} = \frac{h^3}{12}\frac{C_{33}(C_{12} + 2C_{66}) - C_{13}C_{23}}{C_{33}}, \quad I_0 = \varrho h.$$

The coefficients D_{11}, D_{22}, D_{12} and D_{66} are frequently called anisotropic flexural rigidities of the plate. In the case of an isotropic plate, $D_{11} = D_{22} = D_{12} + 2D_{66} = D = h^3 E/(12(1 - v^2))$ is the flexural rigidity of the plate, E is the Young's modulus and v is the Poisson's ratio. Thus, Equation (2.37) becomes the well-known dynamic equation of *Kirchhoff-Love plate*

$$D \left(\frac{\partial^4 w_0}{\partial x^4} + 2\frac{\partial^4 w_0}{\partial x^2 \partial y^2} + \frac{\partial^4 w_0}{\partial y^4} \right) = D\nabla^2\nabla^2 w_0 = q_3(x,y) - \varrho h \frac{\partial^2 w_0}{\partial t^2}. \tag{2.39}$$

The classical laminated plate theory presented in this section is a simplification of the 3D elasticity problem and it allows to describe the wave motion in a composite plate only at the low frequency range, where the wavelengths are larger than the plate thickness [91]. However, due to its simplicity it is frequently used for modelling of Lamb wave propagation in composite structures [106]. Additionally, analysis of the solution of the CLPT-constitutive equations allows to explain many of the phenomena observed for Lamb waves in anisotropic laminated plates.

2.3 Lamb waves in plate-like structures

2.3.1 Wave solution of equation of motion, Bulk waves

The elastodynamic equations of motion in each layer given previously by (2.5) are also known as wave equations for elastic media, i.e. their solution can be represented using the harmonic wave ansatz

$$\mathbf{u}^{(n)}(x,y,z,t) = \hat{\mathbf{u}}^{(n)} e^{i(k_x x + k_y y + k_z z - \omega t)}, \tag{2.40}$$

where $\hat{\mathbf{u}}^{(n)}$ is a polarization (or amplitude) of wave, $\mathbf{k} = (k_x, k_y, k_z)$ is a wave vector (k_i in rad/m). Formula (2.40) represents the solution of the equations of motion as a sum of fundamental, propagating time-harmonic plane waves. Inserting (2.40) into Equations (2.5) yields a 3×3 system, where the eigenvalues give a relation between the wavenumbers k_x, k_y, k_z, an angular frequency $\omega = 2\pi f$ (f is a frequency in Hertz) and material properties [7]. This relation is known as *Christoffel's equation* [24]. Wavenumbers found for the layer n correspond to *bulk waves* in unbounded media of material with properties as in n-th layer of composite plate. In general, all three bulk waves are propagating in the same direction, with different velocities, and with mutually orthogonal polarizations u_i. The wave with the polarization u_i closest to the propagation direction n_i is called quasi-longitudinal (L-wave), the others are called quasi-transverse (TV-wave), vertical shear and horizontal shear waves [117].

2.3.2 Dispersion equation for multilayered plate

The wave motion within a plate is a guided wave motion, representing the superposition of incident and reflected partial bulk waves (TV or L respectively). Reflections occur at the surfaces of the plate. The reflection of incident L (TV) wave from the plate surface generates the TV (L) wave mode [117], the combination of partial reflected

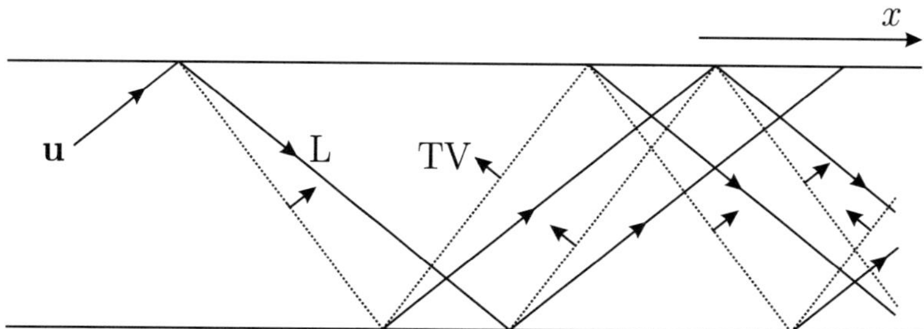

Figure 2.6: Reflection of longitudinal (L) and transverse (TV) bulk waves at plate surfaces. Each reflection of L (TV) generates also the TV (L) wave mode constituent (mode conversion) [117]

waves travelling to infinity results in guided modes propagating along the phase front direction (Figure 2.6) [96]. These *guided* waves are called *Lamb waves* [71]. The equations of guided wave motion are obtained by adjusting the displacements and stresses of bulk waves in particular layers on the boundary conditions (2.9)-(2.10) and interface conditions (2.8). Two approaches are mostly used for obtaining the matrix solution of the problem - transfer [137] and global [65] matrix methods. A detailed review of both methods is given in [82]. Generally, both methods assume that the wavenumbers $k_z^{(n)}(k_x, k_y)$ correspond to standing waves across the waveguide. They are found from Christoffel's equation in each layer in dependence on material parameters, frequency ω and wavenumbers k_x, k_y. A similar way to get a dispersion equation is presented in this work using an integral approach and Green's matrix (for more information see chapter 3). The equality of determinant of the matrix resulting from boundary and interface conditions to zero relates the wavenumbers k_x, k_y of plane waves to thicknesses of individual layers, to their material properties and to frequency ω. This relation, called *dispersion equation*, can be obtained for multilayered structures only in an implicit form [95]

$$\Delta(\omega, k_x, k_y) = 0 \quad \text{or} \quad \Delta(\omega, k, \gamma) = 0, \tag{2.41}$$

if the wavevector is represented in polar coordinates $k_x = k\cos\gamma$, $k_y = k\sin\gamma$. The dispersion of the waves in (2.41) means the variation of wavenumbers with respect to frequency ω: $k_x(\omega)$ and $k_y(\omega)$. Note that Equation (2.41) or some similar dispersion equations can be obtained using other approaches like strip element method (SEM) [79, 125] or semi-analytical finite element method (SAFE) [47, 48]. Traditionally, dispersion equations are solved for wavenumbers k in dependence on circular frequency ω and propagation direction γ. Already in the simplest case of a single-layered isotropic plate the solutions of dispersion equation can be found only numerically (for more information see section 2.3.6.2). Finally, the solution of the problem (2.5), (2.8), (2.10),

(2.12) is represented by the sum of *Lamb waves* as

$$\mathbf{u}^{(n)}(x, y, z, t) = \hat{\mathbf{u}}^{(n)} e^{i(k_x x + k_y y - \omega t)}. \tag{2.42}$$

The representation can be modified by changing variables as

$$\begin{aligned} x &= r \cos \varphi, \quad y = r \sin \varphi, \\ k_x &= k \cos \gamma, \quad k_y = k \sin \gamma, \end{aligned} \tag{2.43}$$

where $r \geq 0$, $\operatorname{Re} k \geq 0$, $\varphi \in [0, 2\pi)$ and $\gamma \in [0, 2\pi)$ into the form of cylindrical waves

$$\mathbf{u}^{(n)}(r, \varphi, z, t) = \hat{\mathbf{u}}^{(n)} e^{i(kr \cos(\gamma - \varphi) - \omega t)}. \tag{2.44}$$

Note that for generating analytical dispersion curves and mode shapes for various configurations of composite materials, several computationally efficient numerical routines have been implemented in the commercial software Disperse [102].

2.3.3 Dimensionless form

Solving the elastodynamic problems, it should be taken into account that in most systems of units the elastic constants are large numbers, typically of the order of 10^{11} in Pa. This may cause numerical difficulties, which can be avoided by dividing all the elastic constants by a typical value of the elastic constants. To get the problem in a dimensionless form suitable for numerical computations, also the density and the thickness of the composite are normalized. In this work, unless otherwise stated, the stiffness in computations are normalized by E_0 - some typical value of stiffness, the density is normalized by some typical value ϱ_0 and the thickness by the whole thickness of laminate h. It yields the dimensionless constants of the model

$$\overline{C}_{ij} = C_{ij}/E_0, \quad \overline{\varrho} = \varrho/\varrho_0, \quad \overline{h} = 1. \tag{2.45}$$

Moreover, the choice of three normalizing parameteres gives a dimensionless form of time t, frequency f, velocity c and wavenumber variables

$$\overline{t} = tc_0/h, \quad \overline{f} = fh/c_0, \quad \overline{c} = c/c_0, \quad \overline{k} = kh, \tag{2.46}$$

where $c_0 = \sqrt{E_0/\varrho_0}$. Note that whereas many authors use for values of E_0 and ϱ_0 the in-plane shear modulus G_{xy} and density ϱ of the constitutive material of laminate (or one of the constitutive materials in case of hybrid laminates), in this thesis the following values are taken:

$$E_0 = 10^{11} \text{ Pa}, \quad \varrho_0 = 10^3 \text{ kg/m}^3. \tag{2.47}$$

Besides the dimensionless frequency $\overline{f}$, the value of *frequency-thickness* product $f \cdot h = \overline{f} c_0$ (or simply frequency-thickness) is considered in this thesis.

2.3.4 Lamb wave modes

Lamb proved theoretically that under certain conditions, a finite number of wave modes can propagate independently in a plate [71]. This means that the total response of a system is splitted into superposed individual wave solutions. Moreover, in case of plate symmetry with respect to mid-plane and orthotropic or higher symmetric properties of all laminae[1], the motion in the plate can be devided into two types: symmetric and antisymmetric. Considering for simplicity the coordinate system with reference plane $z = 0$ equal to mid-plane of the plate that the symmetric motion satisfies conditions $u_x(z) = u_x(-z)$, $u_y(z) = u_y(-z)$ and $u_z(z) = -u_z(-z)$, whereas the antisymmetric motion satisfies conditions $u_x(z) = -u_x(-z)$, $u_y(z) = -u_y(-z)$ and $u_z(z) = u_z(-z)$. From the symmetric wave motion and stress-free boundary conditions (2.9), (2.10) it follows [149] that the in-plane stresses and the out-of-plane displacement are zero at the mid-plane $z = 0$:

$$\left(\sigma_{xz}, \sigma_{yz}, u_z\right)\big|_{z=0} = (0,0,0). \tag{2.48}$$

Similarly, an antisymmetric motion does not produce vertical stress and in-plane displacements at the mid-plane:

$$\left(u_x, u_y, \sigma_z\right)\big|_{z=0} = (0,0,0). \tag{2.49}$$

Due to the symmetry of the plate, only the upper (or lower) half of the plate with respect to a mid-plane may be considered, which allows to reduce the compuational time for solving the dispersion equation (2.41). Applying for the half-plate the corresponding boundary conditions (2.48) for symmetric motion and (2.49) for antisymmetric motion at the mid-plane of the plate $z = 0$ yields two independent dispersion equations

$$\Delta_S(\omega, k, \gamma) = 0 \tag{2.50}$$

and

$$\Delta_A(\omega, k, \gamma) = 0. \tag{2.51}$$

The solutions of dispersion equations are continuos functions of their variables. Moreover, if no principal directions for the composite exist[2], the surfaces of real wavenumbers $k_s(\omega, \gamma)$ satisfying the dispersion equation for symmetric motion (2.50) do not intersect each other. The same applies to the real solutions $k_a(\omega, \gamma)$ of dispersion equation for antisymmetric motion (2.51). The corresponding waves with wavenumbers $k_s(\omega, \gamma)$ and $k_a(\omega, \gamma)$ are called *symmetric* and *antisymmetric* Lamb wave modes. However, the wavenumber surfaces of different families of waves can cross each other [117] and solving the two dispersion equations (2.50), (2.51) instead of one dispersion equation allows to avoid the numerical instabilities due to the intersection points and to distinguish the propagation modes uniquely.

---

[1] Already in case of monoclinic properties, the decoupling of two types of motion is not possible [95].

[2] Principal directions occur in case of single-layered material (e.g. unidirectional composite) with orthotropic or higher symmetry [142].

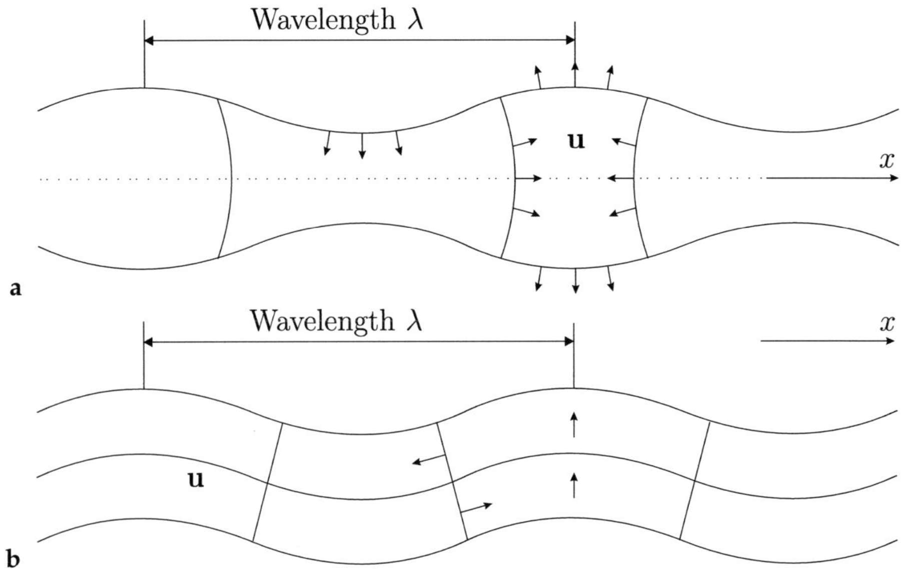

Figure 2.7: Mode shapes of symmetric (a) and antisymmetric (b) fundamental Lamb wave modes. Longitudinal displacements of symmetric (antisymmetric) wave modes are equal (opposite) on either side of the median plane, transverse (vertical) displacements are opposite (equal) [117]

Studying of wavenumber surfaces has shown that at low frequencies two symmetric and one antisymmetric propagating wave modes exist and their normalized wavenumbers $kh \to 0$ as $\omega \to 0$. These three wave modes are usually called *fundamental* wave modes and have different main polarization directions for wave excitation in the x-direction. One of the two symmetric modes named qS_0 is polarized mainly in longitudinal (x) direction (*quasi-longitudinal mode*), the other one named qSH_0 (*quasi-shear horizontal mode*) is polarized mainly in the in-plane direction y perpendicular to propagation direction (x) [111]. The fundamental antisymmetric wave (sometimes termed as *shear vertical* (SV) mode) named A_0 is polarized at low frequencies mainly in vertical (z) direction to the plate surface and is also called *transverse*, *flexural* or *bending* mode. Note that the prefix "quasi" is omitted in the following for wave modes symmetric with respect to mid-plane laminates, i.e. the following notation is used: $S_0 = qS_0$ and $SH_0 = qSH_0$. However, for non-symmetric laminates the notation with prefix "quasi" will further be used.

The mode shapes of fundamental S_0 and A_0 modes are presented schematically in xz-plane in Figure 2.7a and Figure 2.7b respectively. Note that the fundamental wave modes propagate at all frequencies. However, their polarization properties are true only at low frequency range, and at higher frequencies the direction of particle motion of each wave is more complex and as a rule does not coincide with coordinate axes [117].

Due to the propagation at the whole frequency range, the fundamental Lamb modes present a big potential for structural health monitoring. Besides the fundamental wave modes, other waves with *normal dispersion*[1], called *high frequency* or *nonfundamental* wave modes have real wavenumbers only for frequencies above the so called *cut-off frequencies* [75]. Below the cut-off frequency of the mode, the wavenumbers are complex or pure imaginary, and the mode decays exponentially in the waveguide direction, i.e. the mode is non-propagating. In the following the *nonfundamental wave mode* is understood to be a wave mode with real wavenumbers only above its cut-off frequency. The nonfundamental wave modes are denoted similar to fundamental modes according to their polarization at low values of normalized wavenumbers $kh \ll 1$, which occur at frequencies slightly above the cut-off frequency. The notation S_m is used for symmetric longitudinal waves, A_m stands for antisymmetric transverse waves, and SH_m for shear horizontal waves. Note that the shear horizontal waves with even indices are symmetric and with odd indices are antisymmetric. As in the case of fundamental modes, nonfundamental modes exhibit no perfect polarization at frequencies sufficiently above the cut-off frequency.

Remark 2.1 *Notations introduced in this section for Lamb waves in composite structures are valid only in case of symmetric properties of the composite with respect to mid-plane and symmetry class of single layer not lower than orthotropic. This classification is based on the decoupling of symmetric and antisymmetric motions. However, in case of non-symmetric composite plates or lower symmetries of layers, the symmetric and antisymmetric modes are coupled and two independent dispersion equations for both types of wave motion cannot be obtained. In this case Lamb wave modes are called quasi-antisymmetric qA_m and quasi-longitudinal qS_m.*

2.3.5 Dispersion properties of elastic waves

Besides wavenumbers of Lamb waves, also other physical values traditionally used in acoustics can be calculated for Lamb waves. The wavelengths of Lamb waves are calculated as

$$\lambda(\omega,\gamma) = \frac{2\pi}{k(\omega,\gamma)}. \tag{2.52}$$

Another quantity, the slowness, i.e. the quotinent of wavenumbers to frequency, is frequently analysed with respect to Lamb waves:

$$s(\omega,\gamma) = \frac{k(\omega,\gamma)}{\omega}. \tag{2.53}$$

---

[1]The definitions of normal and abnormal dispersions is given in sections 3.1.3, 4.3.1

In practical applications the values of the phase velocity c_p and group velocity c_g of waves are more convenient:

$$c_p(\omega,\gamma) = \frac{\omega}{k(\omega,\gamma)}, \quad c_g(\omega,\gamma) = \frac{d\omega}{dk(\omega,\gamma)}. \tag{2.54}$$

According to the notations used in [79], the surfaces corresponding to phase $c_p(\omega,\gamma)$ and group $c_g(\omega,\gamma)$ velocities are named phase velocity surface (PVS) and group velocity surface (GVS). The PVS and GVS represent the admissible phase and energy velocities of plane wave modes in the propagation direction. Comparing the phase and group velocities of Lamb waves with those of bulk waves it is concluded that both velocities vary with frequency and phase front direction as well [96].

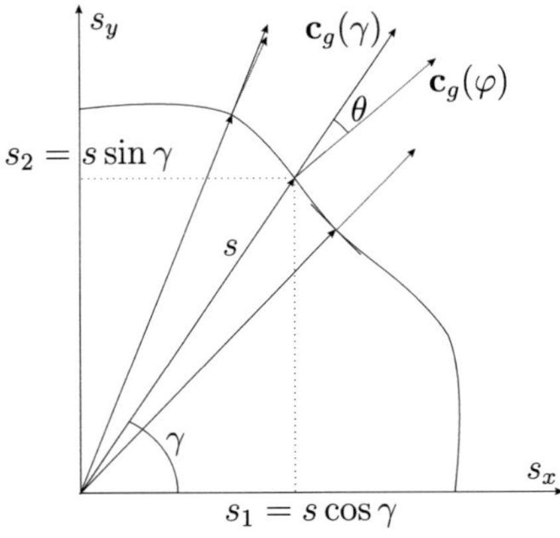

Figure 2.8: Definition of vector of group velocity of observed waves [96]

However, PVS and GVS describe only the velocities of incident waves in a plate. The directions in which these waves can be observed in a plate are given by the normal to the slowness surface [79] (Figure 2.8). Generally due to the anisotropy of the structure this direction does not coincide[1] with the direction of incident waves in a plate. The surfaces of phase and group velocities in dependence on an observation direction φ (2.43) are also called *wave surfaces* [79] or *wave curves* [149] and can be obtained as all points traced by the tip of the corresponding velocity vectors drawn from a fixed origin as the propagation direction is varied [79]. The phase wave surface (PWS) and the group wave surface (GWS) define the corresponding characteristics of wave fronts observed in a composite plate in case of wave excitation by a point source. The simplest way to calculate GWS is given in [149], the wave surfaces are calculated for a fixed frequency $\omega = \omega_0$ in Cartesian coordinates in an implicit form as

$$\begin{pmatrix} c_{gx} \\ c_{gy} \end{pmatrix} = c_g(\omega_0,\gamma) \begin{pmatrix} \cos\gamma & -\sin\gamma \\ \sin\gamma & \cos\gamma \end{pmatrix} \begin{pmatrix} 1 \\ -k'_\gamma(\omega_0,\gamma)/k(\omega_0,\gamma) \end{pmatrix} \tag{2.55}$$

---

[1]Except of isotropic properties of all layers (see section 2.3.6.2).

where $k'_\gamma(\omega_0, \gamma) = dk(\omega_0, \gamma)/d\gamma$. Corresponding phase wave surfaces for all propagating wave modes can be obtained by replacing the group velocity $c_g(\gamma, \omega_0)$ by the phase velocity $c_p(\omega_0, \gamma)$. The values of phase and group velocities are appropriate for identification of propagation modes in an experimental setup by comparison of measured and theoretical values [149].

Using an implicit formulation (2.55) the observation direction φ for Lamb waves excited in direction γ is found to be

$$\varphi \;=\; \arctan\frac{c_{gy}}{c_{gx}} = \arctan\frac{c_{py}}{c_{px}} \tag{2.56}$$

$$= \; \arctan\left(\frac{\sin\gamma - \cos\gamma \cdot k'_\gamma(\omega_0, \gamma)/k(\omega_0, \gamma)}{\cos\gamma + \sin\gamma \cdot k'_\gamma(\omega_0, \gamma)/k(\omega_0, \gamma)}\right).$$

Letting $k'_\gamma(\omega_0, \gamma)/k(\omega_0, \gamma) = \tan(\theta(\gamma))$ it follows that

$$\varphi = \arctan\left(\frac{\tan\gamma - \tan(\theta(\gamma))}{1 + \tan\gamma\,\tan(\theta(\gamma))}\right) = \gamma - \theta(\gamma). \tag{2.57}$$

The deviation angle of the observed waves in direction φ from the propagation direction γ (direction of the wave vector) is known as the skew angle or (beam) *steering angle* [120]

$$\theta = \gamma - \varphi. \tag{2.58}$$

Besides of such an influence of the anisotropy as the deviation of waves from the incident direction, the wave surfaces corresponding to a single wave mode can be also multifolded, i.e. in a single direction more than one wave of the same type can be observed. This occurs if the Lamb wave mode presents inflexion points on its slowness surface, which implies characteristic cusps on its associated wave surface [96]. In directions near to the cusps, so called *caustics*, two (or more) waves corresponding to the wave mode are observed with nearly the same velocities. The energy carried out by this wave mode concentrates near to caustics. This phenomenon of energy focussing is well-known for bulk waves in anisotropic media [117]. In more detail the focussing of Lamb waves will be investigated within the next chapters.

Note also that the typical relation between slowness $\mathbf{s} = (s_x, s_y)$ and group velocity $\mathbf{c}_g = (c_{gx}, c_{gy})$ vectors valid for bulk waves $(\mathbf{s}, \mathbf{c}_g) = 1$ or $c_p = c_g\cos\theta$ (where θ is the steering angle) [7] does not hold for Lamb waves because of the dispersive nature of Lamb waves. It can be obtained as [96]

$$(\mathbf{s}(\omega, \gamma), \mathbf{c}_g(\omega, \gamma)) = 1 + \frac{\omega\partial_\omega\Delta(\omega, s, \gamma)}{\mathbf{s}\partial_{\mathbf{s}}\Delta(\omega, s, \gamma) - \omega\partial_\omega\Delta(\omega, s, \gamma)}, \tag{2.59}$$

where

$$\partial_\omega\Delta(\omega, s, \gamma) = \frac{\partial\Delta(\omega, s, \gamma)}{\partial\omega}, \quad \partial_{\mathbf{s}}\Delta(\omega, s, \gamma) = \left(\frac{\partial\Delta(\omega, s, \gamma)}{\partial s_x}, \frac{\partial\Delta(\omega, s, \gamma)}{\partial s_y}\right). \tag{2.60}$$

It follows that the group velocity of the guided wave modes can be greater or lower than its associated phase velocity.

2.3.6 Lamb waves in single-layered structures

The free propagation of Lamb waves in anisotropic structures is studied in many works. However, most of the results are obtained for isotropic or orthotropic single-layered plates. Single-layered plates are obviously symmetric with respect to mid-plane and therefore the properties of wave modes present the partial case of Lamb waves in multilayered structures. Some of the most important facts about the propagation of waves for such partial cases are summarized in this section.

2.3.6.1 Lamb waves in triclinic and orthotropic plates

The dispersion equation of Lamb waves in a triclinic (fully anisotropic) plate can be obtained in an analytical but very complicated form [95]. Moreover, the wave motion in a triclinic plate is coupled and all wave modes are only quasi-symmetric or quasi-antisymmetric. Hence, in industrial applications the single-layered plates are usually orthotropic or transversally-isotropic (unidirectional composites). It is sufficient to analyse the properties of waves in orthotropic plates, because waves in the plates with higher symmetries present only a partial case of an orthotropic plate [142].

Lamb waves in an orthotropic plate at low frequencies can be decoupled into symmetric and antisymmetric wave modes. Moreover, due to the existence of principal axes in such a structure, the symmetric Lamb modes S_m and SH_m are decoupled in principal directions [96]. This decoupling in such a propagation direction is due to the decoupling of equations of motion into two equations of motion in the saggital plane xz and one equation of motion for pure horizontal shear motion. This decoupling into partial motions allows the intersection of dispersion surfaces corresponding to symmetric modes, which is not admitted if no principal directions occur. For wave propagation in direction not coinciding with any of the principal axes directions, the corresponding equations take a form similar to the monoclinic symmetry case and both symmetric wave modes are coupled [95]. Moreover, another interesting phenomenon can occur in an orthotropic plate. For example, in some situations the modes S_0 and SH_0 polarized for the principal direction of $0°$ mostly in longitudinal and transverse in-plane directions respectively can swap their main polarization while rotating the propagation direction until another principal direction of $90°$ [129]. The wave modes in such a case are labeled concerning their polarization in $\gamma = 0°$ principal direction.

Due to the orthotropy of the plate, the wavenumbers of Lamb wave modes are dependent on the angle γ and the value $k'_\gamma(\omega_0, \gamma)$ is commonly nonzero. This yields that the steering of waves $\theta(\gamma)$ differs from zero. Nevertheless, in principal directions all wave modes have no steering [96]. The velocities of waves depend also on propagation direction, and velocities of S_0 and A_0 reach their maximum values in directions which correspond to the stiffer values of C_{ijkl}. Their minimum values are reached in the direction of another principal axis at an angle of $90°$. For more information on Lamb waves in single-layered plates the readers are referred to the work [96].

2.3.6.2 Lamb waves in an isotropic plate

The isotropic symmetry of the plate represents the simplest case of Lamb wave motion in plate-like structures. Each of the directions can be considered as a principal axis direction. It follows that compared with the orthotropic case, the symmetric shear horizontal wave motion is completely decoupled from the longitudinal motion in all propagation directions. The corresponding dispersion equations for waves of three possible polarizations can be derived in a simple analytical implicit form, given firstly by Viktorov in [143]:

$$\frac{\tan(\sigma_s h/2)}{\tan(\sigma_l h/2)} = - \left(\frac{4k_1^2 \sigma_s \sigma_l}{(\sigma_s^2 - k^2)^2} \right)^{\pm 1} \tag{2.61}$$

where the plus "+" and minus "−" signs correspond to Lamb wave equation for longitudinal (S_m) waves and flexural (A_m) waves, $\sigma_l^2 = \omega^2/c_l^2 - k^2$. $\sigma_s^2 = \omega^2/c_s^2 - k^2$ and $c_l = \sqrt{C_{11}/\varrho}$, $c_s = \sqrt{C_{44}/\varrho}$ are the velocities of longitudinal and transverse waves in unbounded isotropic medium respectively, $k = \sqrt{k_x^2 + k_y^2}$. The properties of the structure as well as the wavenumbers of Lamb modes are independent of the propagation direction γ. Also the dispersion equation of the isotropic plate (2.61) is frequently called *Rayleigh-Lamb frequency equation*.

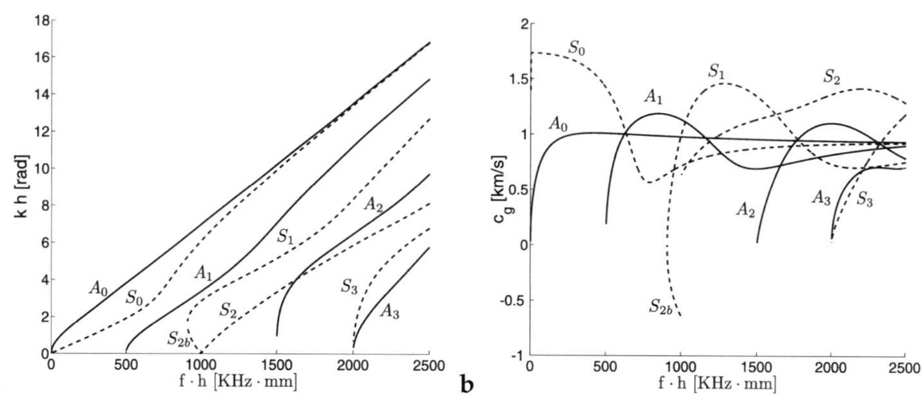

Figure 2.9: Dispersion curves in a steel plate. The wavenumbers (a) and group velocities (b) of wave modes propagating at frequency-thickness product values lower than $f \cdot h = 2500$ MHz · mm

Lamb waves in an isotropic plate have some characteristic properties. Due to the full decoupling of particle motions in a plate, the symmetric wave modes compared to the case of lower material symmetries are called pure longitudinal S_m and pure shear horizontal SH_m. Since in (2.61) $k'_\gamma(\omega_0, \gamma) = 0$ the corresponding steering angle (2.57) of Lamb waves is zero. This means that the direction of the wave front φ coincides with the direction of the wave vector γ.

Lamb waves in isotropic plates are independent of the propagation direction but are still dispersive. The dispersion curves for a steel plate (Poisson's ratio $\nu = 1/3$) are plotted in Figure 2.9. As illustrated in Figure 2.9, at low frequencies, the phase velocity of S_0 tends to the velocity of the longitudinal wave c_l. As $\omega h \to \infty$ ($\omega \to \infty$ or $h \to \infty$), the plate becomes a half-space. The wave velocities of fundamental wave modes S_0 and A_0 tend to the velocity of the Rayleigh surface wave, the velocities of all non-fundamental Lamb modes tend to the shear wave velocity [117]. The cut-off frequency of the first higher-order wave mode A_1 is $f \cdot h = 500$ KHz $\cdot$ mm. The backward wave mode S_{2b} (i.e. wave mode with abnormal dispersion) appears in the range $f \cdot h \in [920, 1000]$ (in KHz $\cdot$ mm).

2.3.6.3 Bending waves in orthotropic and isotropic plates

The harmonic plane wave ansatz (2.40) used for elastodynamic equation obey also the equations of plate theories (Appendix A.6 and A.7). The corresponding wavenumbers k are found in dependence on wave propagation direction γ and frequency ω from the corresponding plate theory dispersion equation. If the composite plate is symmetric with respect to mid-plane, the matrix $\mathbf{B} = \mathbf{0}$ and the symmetric and antisymmetric motions are decoupled. In this section some of the properties of antisymmetric waves are discussed for the example of a plate modelled using CLPT. The corresponding dispersion equation for antisymmetric motion takes the form

$$k^4 \left(D_{11} \cos^4 \gamma + 2(D_{12} + 2D_{66}) \cos^2 \gamma \sin^2 \gamma + D_{22} \sin^4 \gamma \right) - I_0 \omega^2 = 0. \tag{2.62}$$

There is one positive real number solution of this equation, which is called a pure *bending* wave mode. It is found to be

$$k(\gamma, \omega) = \sqrt{\omega} \sqrt[4]{\frac{I_0}{\cos^4 \gamma D_{11} + 2(D_{12} + 2D_{66}) \cos^2 \gamma \sin^2 \gamma + D_{22} \sin^4 \gamma}} \tag{2.63}$$

$$= \sqrt{\omega} k_0(\gamma).$$

The negative real roots correspond to the same wave propagating in an opposite direction. The pure imaginary roots of Equation (2.62) correspond to non-propagating waves and do not transport energy.

The values of phase velocity, slowness and group velocity for the bending wave mode are obtained as

$$
\begin{aligned}
c_p(\gamma,\omega) &= \frac{\omega}{k} = \frac{\omega}{\sqrt{\omega}k_0(\gamma)} = \frac{\sqrt{\omega}}{k_0(\gamma)}, \\
s(\gamma,\omega) &= \frac{k}{\omega} = \frac{k_0(\gamma)}{\sqrt{\omega}}, \\
c_g(\gamma,\omega) &= \frac{d\omega}{dk} = \left(\frac{dk}{d\omega}\right)^{-1} = 2\frac{\sqrt{\omega}}{k_0(\gamma)}.
\end{aligned}
\tag{2.64}
$$

Note that CLPT is valid only for describing the dispersive solutions of guided wave modes at frequencies at which the wavelengths are larger than the plate thickness [87].

Nevertheless, the bending wave mode in anisotropic plates represent some interesting results about the influence of the anisotropy on fundamental wave modes. For example, analyzing

$$
\frac{dk}{d\gamma} = \sin\gamma\cos\gamma\frac{\sin^2\gamma(D_{12}+2D_{66}-D_{22})+\cos^2\gamma(D_{11}-D_{12}-2D_{66})}{\cos^4\gamma D_{11}+2(D_{12}+2D_{66})\cos^2\gamma\sin^2\gamma+D_{22}\sin^4\gamma},
\tag{2.65}
$$

it can be concluded that the steering angle (2.57) for bending mode does not depend on the frequency. The steering angles for some propagation directions can be calculated in an analytical form: $\theta = 0°$ for $\gamma = 0°$ and $\gamma = 90°$, $\theta = \pi/4 + \arctan(D_{22}-D_{11})/(D_{11}+D_{22}+2D_{12}+4D_{66})$ for $\gamma = \pi/4$. It follows that as in case of analytical modelling for elastodynamic equations, the steering angles are equal to zero in principal directions.

The example of the wavenumber curves for the bending mode in a unidirectional plate manufactured of IM7-Cycom-977 (material properties are given in Table A.1 in Appendix A.9) is shown in Figure 2.10. Figure 2.10a shows the frequency dependence of dispersion curves for directions $\gamma = 0°$ (straight line) and $\gamma = 45°$ (dotted line). Figure 2.10b shows implicitly the dependence on φ of group velocity of bending wave observed in the plate (GWS) for a fixed frequency-thickness $f \cdot h = 11$ KHz $\cdot$ mm.

In the isotropic case Equation (2.62) can be simplified due to the relation $D_{11} = D_{22} = D_{12} + 2D_{66} = D$. The wavenumber of bending mode for a Kirchhoff-Love plate similar to the wavenumbers of Lamb waves in isotropic plate does not depend on the propagation angle γ

$$
k(\gamma,\omega) = k(\omega) = \sqrt{\omega}\,\sqrt[4]{\frac{I_0}{D}},
\tag{2.66}
$$

i.e. $k_0(\gamma) \equiv k_0 = \sqrt{I_0/D}$. It follows that the steering angle is equal to zero and waves are observed in the same directions as they are excited.

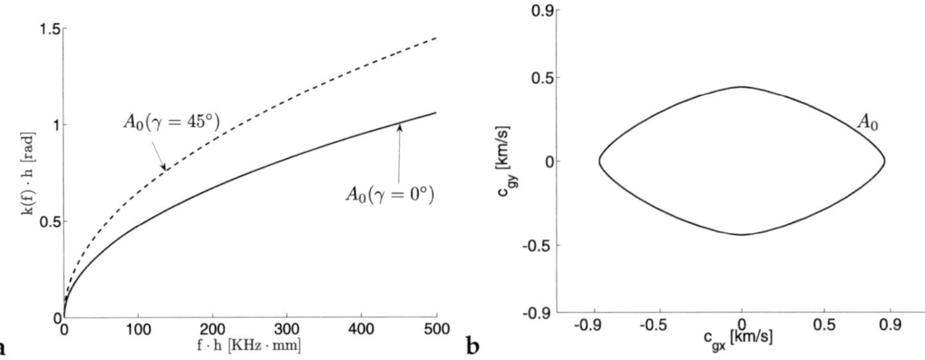

Figure 2.10: Dispersion curves in the IM7-Cycom-977 unidirectional plate. (a) Wavenumbers of bending mode computed using CLPT in directions $\gamma = 0°$ ("−") and $\gamma = 45°$ ("−−"). (b) GWS (2.55) of bending mode at fixed frequency-thickness 11 KHz · mm

2.4 Modelling of surface-bonded wave excitation sources

2.4.1 Piezoelectric wafers for wave excitation and wave sensing

Waves in elastic structures can be excited on different ways. While for NDT excitation devices can be used, which are not coupled (or weakly) with the structure, for SHM the wave actuators need to be integrated into the structure or mounted on its surface. For example, the perspex wedgecoupled angle-adjustable ultrasonic probes (ultrasonic transducers) frequently used in NDT are normally non-negligible due to their weight and sizes because the properties of the structure can be affected by the transducers considerably. Hence, they are not suitable for using in SHM techniques [132].

An alternative excitation method is presented by piezoelectric lead zirconate titanate (PZT) wafer/elements, which excite Lamb waves directly through the electromechanical coupling effect in a piezoelement: "The piezoelectric effect creates a mechanical stress in a piezoelement, when an electric field (a voltage) is applied across it, or conversely, it creates a voltage when a mechanical stress is applied" [75]. If such a voltage is oscillating, it produces propagating oscillatory waves due to the in-plane strain coupling between PZT wafer and the structure. Note that in contrast to PZT elements, also called Piezoelectric Wafer Active Sensors (PWAS), the ultrasonic transducers are characterized by the displacement coupling [33]. Due to the electromechanical coupling in PZT elements, the wafers can also be used as sensors for measuring of propagating waves. Hence, the PWAS are most suitable for using as built-in or surface-mounted wave actuators and sensors because of their sizes, light weight and low cost [132].

For an activation of diagnosing Lamb waves for damage identification by piezoelectrical wafers, the appropriate wave mode, wave form, wave magnitude and wavelength should be chosen. These parameters can be controlled in excitation using PWAS by selecting the appropriate actuator shape and dimensions and by choosing the driving electric voltage signal. Another important factor which influences the wave propagation [25, 32] is the adhesive layer (layer of glue) used for the mounting of the piezoelectric patch on the surface of a host structure. In the following the modelling techniques for the interaction between piezoelectric wafers, bonding layer and laminated composite plate as well as the properties of the most commonly used wafers for SHM are briefly described.

2.4.2 Excitation signals

The analysis of waves in plate-like structures becomes complicated due to their dispersive nature. In order to level a wave dispersion and enhance the sensitivity of waves to damages and simplify subsequent signal processing and interpretation, it is appropriate for practical applications to use excitation signals, the frequency spectrum of which is concentrated near one frequency. Some of transient input signals widely used in practical applications are presented in this section.

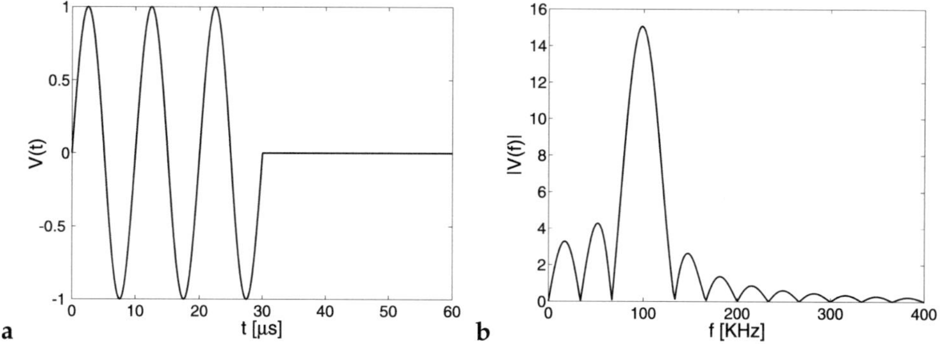

Figure 2.11: Three-cycles-sine excitation signal (a) with central frequency $f_c = 100$ KHz and its frequency spectrum (b)

As a first type of finite excitation pulses the n-cycles sine tone bursts are considered [39]

$$v_n(t) = \begin{cases} \sin \omega_c t, & 0 \leq t \leq nT \\ 0, & t < 0 \text{ or } t \geq nT \end{cases}'$$ (2.67)

where the period of the oscillations $T = 2\pi/\omega_c$ is defined using a central frequency of an excitation signal $\omega_c = 2\pi f_c$. The frequency spectrum of this pulse is found to be

$$V_n(\omega) = \int_0^{nT} v_n(t)\, e^{i\omega t}\, dt = -\frac{\omega_c}{\omega^2 - \omega_c^2}\left(1 - e^{i\omega nT}\right)$$ (2.68)

As observed from (2.68), for an increasing number of cycles n the spectrum becomes more and more concentrated near a frequency f_c. Such an excitation signal with 3 sine cycles with a central frequency $f_c = 100$ KHz is presented in Figure 2.11a. Its frequency spectrum is plotted in Figure 2.11b. A concentration of the signal spectrum near to its central frequency $f_c = 100$ KHz is clearly observed here, so the wave dispersion can be considerably reduced [132]. The main disavantage in using such driving signals lies in a possible significant contribution of waves actuated at high frequencies, so that in spite of using relatively low central frequency also high frequency effects should be taken into account.

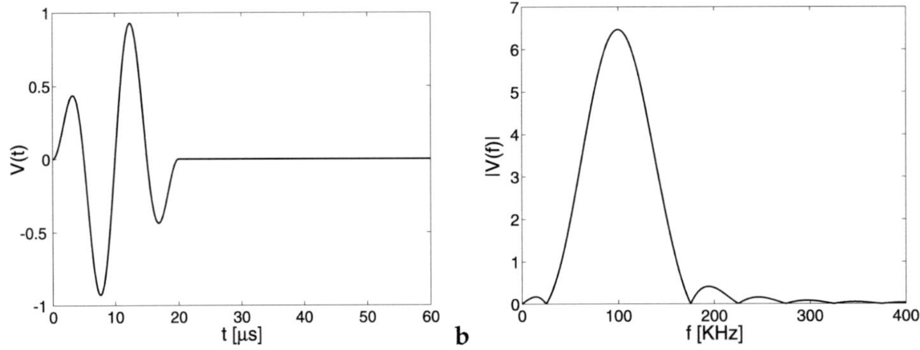

Figure 2.12: Two-cycles-sine-windowed excitation signal (a) with central frequency $f_c = 100$ KHz and its frequency spectrum (b)

As another frequently used driving pulse a two-cycle sine-windowed sine excitation signal is considered [41]

$$V(t) = \begin{cases} \sin \omega_c t \sin \dfrac{\omega_c}{4} t, & 0 \leq t \leq 2T \\ 0, & t < 0 \text{ or } t \geq 2T \end{cases}.$$ (2.69)

As for an n-cycles sine toneburst, the corresponding frequency spectrum is also concentrated near an excitation frequency ω_c:

$$V(\omega) = \int_0^{2T} v(t)\, e^{i\omega t}\, dt = -\frac{128 i \omega \omega_c^2}{256\omega^4 - 544\omega^2 \omega_c^2 + 225\omega_c^4}\left(1 + e^{i\omega 2T}\right).$$ (2.70)

An example of such an actuation signal with central frequency of 100 KHz and corresponding frequency spectrum are plotted in Figure 2.12. Comparing this spectrum with the previous one for *n*-cycles sine tone burst (Figure 2.11) it can be concluded that amplitudes of high- and low- frequency waves are much smaller than for a main frequency band. As a result only this main frequency band is needed to be analysed.

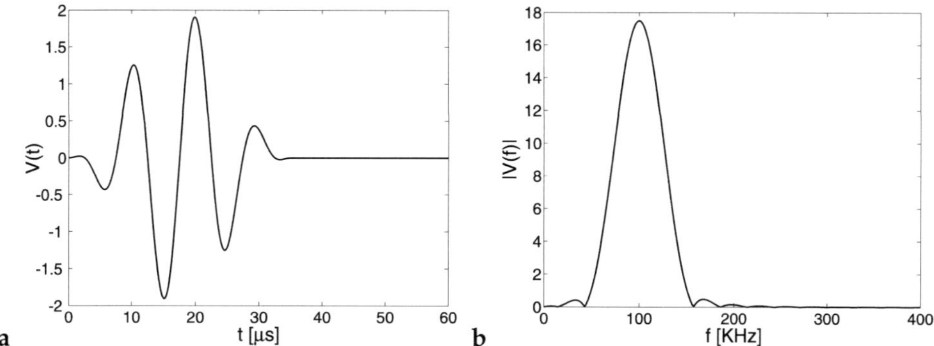

Figure 2.13: 3.5-cycles Hann-modulated excitation signal (a) with central frequency $f_c = 100$ KHz and its frequency spectrum (b)

A concentration of waves can be achieved also in the case of a 3.5 Hann-modulated toneburst [121]

$$v(t) = \begin{cases} \left(1 - \cos\left(\frac{2}{7}\omega_c t\right)\right) \cos\omega_c t, & 0 \leq t \leq 3.5T \\ 0, & t < 0 \text{ or } t \geq 3.5T \end{cases}. \tag{2.71}$$

The corresponding frequency spectrum is obtained in the form

$$V(\omega) = \int_0^{3.5T} v(t)\,e^{i\omega t}\,dt \tag{2.72}$$

$$= \frac{4i\omega\omega_c^2\left(49\omega^2 + 143\omega_c^2\right)}{2401\omega^6 - 7595\omega^4\omega_c^2 + 7219\omega^2\omega_c^4 - 2025\omega_c^6}\left(1 + e^{i\omega 3.5T}\right).$$

This excitation pulse with central frequency 100 KHz and corresponding frequency spectrum are plotted in Figure 2.13. It seems to be a more appropriate for practical applications because of the concentrations of the signal near a central frequency, and the contribution of low- and high- frequency bands is negligible.

Similarly as shown in [77], a transient five-peak input voltage signal applied on the PZT actuator is given according to the formula

$$v(t) = \begin{cases} \left(1 - \cos\left(\frac{1}{5}\omega_c t\right)\right) \sin\omega_c t, & 0 \leq t \leq 5T \\ 0, & t < 0 \text{ or } t \geq 5T \end{cases}. \tag{2.73}$$

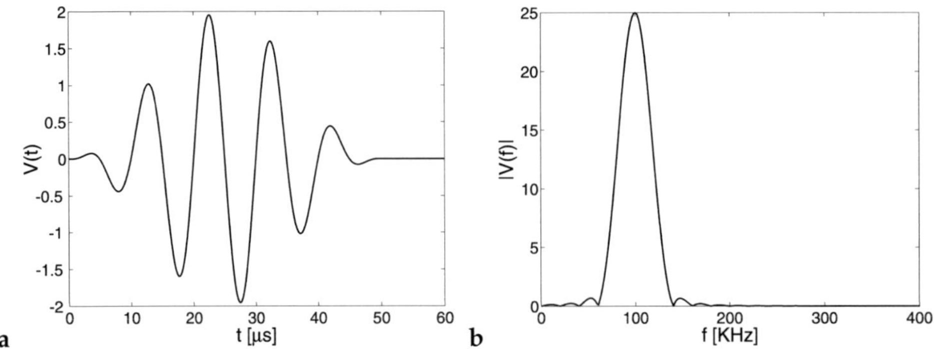

Figure 2.14: The five-peak Hann-modulated excitation signal (a) with central frequency $f_c = 100$ KHz and its frequency spectrum (b)

The frequency spectrum is then obtained as

$$V(\omega) = \int_0^{5T} v(t)\, e^{i\omega t}\, dt = \frac{3\omega_c^3 \left(25\omega^2 + 8\omega_c^2\right)}{625\omega^6 - 1925\omega^4\omega_c^2 + 1876\omega^2\omega_c^4 - 576\omega_c^6} \left(1 - e^{i\omega 5T}\right). \quad (2.74)$$

With increasing number of cycles the signal becomes more and more concentrated near the central frequency allowing the consideration of the excitation as nearly steady-state harmonic. The corresponding signal and its frequency spectrum for a central frequency of 100 KHz are presented in Figure 2.14. However, this signal has a long duration of the wave package, which complicates the signal processing and the following usage of the measured data for damage identification due to overlapping of different wave components [132].

Summarizing this section, the electric driving voltages can be controlled by selecting appropriate excitation frequency, bandwidth, cycle number depending on the properties of waves needed to be excited.

2.4.3 Approaches for modelling piezo-structure interaction

The PZT wafers under the action of the electric voltage produce an interfacial stress between the actuator and the structure. The corresponding stresses are usually unknown and depend on the electric field applied on the surface, the shape of the actuator, the excitation frequency (for the case of harmonic excitation), on the properties of the plate under excitation and on the way the wafer is mounted on the surface. Usually the actuators and sensors are bonded onto the host structure using a thin adhesive layer (glue).

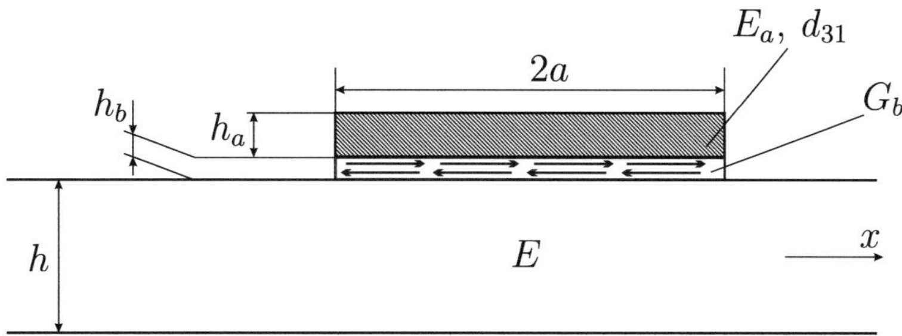

Figure 2.15: Transmission of mechanical stresses from the PZT wafer to the host structure through the adhesive layer [33]

The interaction between the piezoelectric element and host structure is a complex process that takes into account the dynamics of the piezoactuators and sensors, the dynamics of adhesive layers and the dynamics of the plate itself, i.e. this is the *coupled problem*. A good review of numerical techniques used for the modelling of the contact dynamics of piezoelectric actuators and elastic host structures is given in [54]. The dynamic coupled contact problem of Lamb wave excitation by PZT wafers was considered in [37, 127] for actuators bonded on one side of the host structure and in [42] on its both sides. However, the effect of the bonding layer was not considered, i.e. *ideal bonding* was assumed. The problem was reduced to a system of differential-integral equations obtained using the equations of motion of the piezoelectric element, the elastodynamic equations of motion of the host structure, the conditions of ideal bonding between the plate and the piezoelectric element, and boundary conditions. Such a formulation as a system of differential-integral equations was used for the investigation of many contact problems in works [10, 100, 128] and is referred in detail in [146]. However, the dynamic coupled contact problem of Lamb wave excitation by PZT wafers was solved in [37, 127] only for a *plane strain* case, i.e. assuming the independence of the load and host structure properties on one of the in-plane coordinates, e.g. y, which leads to zero strains: $\varepsilon_{xy} = \varepsilon_{yy} = \varepsilon_{yz} = 0$, and considering only single-layered isotropic host material. In [27, 92] Moulin proposed a hybrid approach to model integrated Lamb wave generation with piezo-acuators, obtaining the interfacial stresses due to the piezoelectric element numerically as prescribed excitation and describing the resulting wave propagation in the host structure by the mode expansion method. In [99] the guided wave generation, propagation and reception in an isotropic plate with bonded PZT (lead-zirconate-titanate) wafers is simulated by FE modelling. However, the FE model cannot be applied for large structures, since it operates within spatially restricted discretization [54]. The problem in three-dimensional formulation, taking into account the anisotropy of the properties of the composite under excitation was not investigated in aforementioned works. Other approaches considering also the coupled problem as the finite difference (FD) technique and the finite element method

(FEM) are very time-consuming and not applicable for practical needs. Thus, the problem statement in coupled form is not suitable for a real industrial application at the moment due to its complexity.

Another approach for modelling of interaction between the PZT wafer and an isotropic plate was proposed firstly in [25] for a *plane strain* problem. It is based on the quasi-static modelling, taking also into account the effect of adhesive layer, known as *shear lag effect*, and relying on the independence of contact stresses on the excitation frequency, i.e. the model is *uncoupled*. The piezo-structure interaction through the bonding layer is schematically represented in Figure 2.15. Applying the electric field to a vertically polarized PZT by a voltage given by $V(t)$ and assuming that only axial stresses are produced (due to the piezo-patch flexibility), the in-plane induced strain in the PZT wafer is calculated as follows

$$\varepsilon_{\mathrm{ISA}} = d_{31}\frac{V(t)}{h_a}, \tag{2.75}$$

where d_{31} is the *piezoelectric strain coefficient* (in m/V) describing the coupling between the vertically polarized electric field and in-plane induced strains and h_a is the thickness of the piezo [33]. Assuming stress-free boundary conditions at the edges $x = \pm a$ of the piezopatch, the interface stress in a bonding layer under static and low-frequency actuation of Lamb waves can be found in the form [33]

$$\tau_{xz}(x) = \frac{h_a}{a}\frac{\psi}{\psi+\alpha}E_a\varepsilon_{\mathrm{ISA}}\left(\Gamma a\frac{\sinh\Gamma x}{\cosh\Gamma a}\right) \tag{2.76}$$

for $|x| < a$, where

$$\Gamma^2 = \frac{G_b}{E_a}\frac{1}{h_a h_b}\frac{\alpha+\psi}{\psi} \tag{2.77}$$

with the shear-lag parameter

$$\psi = \frac{Eh}{E_a h_a}. \tag{2.78}$$

G_b is the shear modulus of the bonding (adhesive) layer, h_b the thickness of the adhesive layer, h the thickness of the structure under excitation, E is Young's modulus of the structure, h_a the thickness of PZT wafer and E_a its Young's modulus. Outside the interval $|x| < a$, the surface stress is zero. The parameter α in (2.77) (also known as *modal repartition number*) depends on the stress, strain, and displacement distributions across the plate thickness [25]. The constant distribution of displacements across the plate thickness for the fundamental symmetric mode S_0 at low frequencies gives a value $\alpha = 1$, the linear distribution of displacements across the plate thickness for the fundamental antisymmetric mode A_0 at low frequencies gives a value $\alpha = 3$. Since both wave modes are excited simultaneously during the piezo-structure interaction, the value of this factor is $\alpha = 4$. Note that in work [156] this *classic shear lag solution* was extended to the case of the frequency-dependent amplitude of Lamb waves (nonlinear dependence) and non-fundamental wave modes, which gives an integro-differential equation, which is solved by numerical methods in [122]. For example, the value of α at a value of frequency-thickness product $fh = 780$ KHz (below the first cut-off frequency) for an aluminium plate was found in the referenced work to be $\alpha \approx 5$. However, the contribution of non-propagating evanescent Lamb waves was not considered and the method applied differs from the more common approach used in [127].

The *classic shear lag solution* (2.76) described previously can be simplificated in case of a very thin bonding layer. Assuming $h_b \ll 1$, one obtains $\Gamma \gg 1$, i.e. the shear transfer process becomes very rapid and concentrates over some infinitesimal distances at the ends of the PZT wafer. In the limit, as $h_b \to 0$, i.e. in case of *ideal bonding*, the force is transfered over an infinitesimal region at the edges of the patch[1], and the induced strain is assumed to be given by two concentrated forces applied at its ends (also known as *pin-force model*) [33]

$$\tau_{xz}(x) = a\tau_0 \left[\delta(x-a) - \delta(x+a)\right],\qquad(2.79)$$

where $\tau_0 = (G_b\varepsilon_{\text{ISA}}/h_b\Gamma^2 a^2)$ or $\tau_0 = (G_b\varepsilon_{\text{ISA}}a/h_b\Gamma a \cosh \Gamma a)$. The pin-force model represents the first-order approximation of the piezo-structure interaction and allows to get simple solutions of the wave propagation problem and hence represents a useful tool for SHM.

The pin-force model developed for a two-dimensional problem [25] can also be used for fully 3D problems if the host structure has quasi-isotropic or isotropic properties [132]. Then the interface stress is concentrated on the boundary of the piezoactuator. This model is succesfully used in a various number of works [33, 39, 41, 108] for modelling the interaction between piezopatches and isotropic structures as well as for the case of absence of quasi-isotropy of the structure under excitation [107, 110]. The concrete examples of piezoelectric actuators used in industrial applications and known from the literature with corresponding pin-force models are given below in section 2.4.5. On the contrary, the work [53] takes into account also the effect of the actuator bending and concludes that the waves produced by axis-symmetric piezo-patches in isotropic structures are directional and not axis-symmetric, differently as expected, a result inconsistent with the pin-force model. This is because the decoupled models do not allow to understand the processes occuring in piezo-structure interaction [38]. Moreover, the pin-force model has the following limitations [54]:

- The model is a good approximation only if Young's modulus and thickness of the actuator are small compared to those of the host structure or the bonding layer is very thin and stiff,

---

[1]In reality, the piezopatch transmits the stress to the plate on average at 10% of its length near to its edges [122].

- the model can only provide qualitative estimation about the actuation mechanism for low-frequency cases, which needs to be calibrated by either numerical simulation or experimental testing, and

- piezoelectric resonance effects cannot be captured in the model.

However, the results obtained in works [33, 39, 41, 108, 155] have shown a good coincidence with experimental data, which promise the adequacy of the pin-force model for describing the piezoelecric actuation in many cases. Therefore, in this thesis the pin-force model is used for describing the piezoelectric actuation of elastic Lamb waves in composite materials.

2.4.4 Use of piezo-actuators as sensors

The PZT wafers can develop an output voltage under mechanical deformations due to their elecromechanical properties, so they can also be employed as sensors. Hence, they are also called *Piezoelectric Wafer Active Sensors (PWAS)* [33]. Again, PWASs are the most promising tool for measuring elastic waves for SHM.

In a common case the modelling of the sensing process should take into account the interaction between the piezo patch with propagating Lamb waves, i.e. the problem is coupled [39]. Nevertheless, due to the light weight of piezos and their thinness, under the condition of low-frequency motion of the host structure in absence of an external electric field and assuming the ideal bonding of sensors to the structure, the piezos will undergo only in-plane deformations ε_{xx} and ε_{yy}. Thus, the output voltage V_c captured by PWAS can be computed using the formula[1] [41, 77]

$$V_c = \frac{A_c}{S_c} \iint\limits_{\Omega_c} \left(\varepsilon_{xx} + \varepsilon_{yy} \right) \; dxdy \tag{2.80}$$

where Ω_c corresponds to the sensor contact area, S_c is the area of the domain Ω_c and A_c depends on electrical and mechanical properties of the sensor: $A_c = d_{31} E_a h_a$ [155]. Taking into account the small dimensions of the PZT wafer compared with the structure dimensions, the strains in (2.80) can be assumed to be well approximated by the strain at the center of the wafer $\varepsilon_{cen} \approx \varepsilon_{xx} \approx \varepsilon_{yy}$. Then, the output signal captured by a PWAS can be defined as

$$V_c = 2A_c \varepsilon_{cen}. \tag{2.81}$$

So, the output voltage captured by a PZT wafer used as a sensor is proportional to the central strain of the wafer. However, the use of the more general formula (2.80) makes possible an analysis of the influence of the sensor shape on the Lamb wave

---

[1]Note that in [87] more precise formulas taking into account the anisotropic properties of the piezoactuator are given.

propagation while formula (2.81) allows to get simple but not precise results. Note also that Equations (2.80)-(2.81) are derived for ideal-bonded PWAS. Including the adhesive layer in the analysis it can be concluded that the sensor would develop less voltage across its boundaries and hence would underestimate the measured strain [17]. This is because of the aforementioned shear lag effect, which can be quantified by computing the effective length l_{eff} of the sensor

$$l_{\text{eff}} = 2a \left(1 - \frac{\tanh(2\Gamma a)}{2\Gamma a} \right). \tag{2.82}$$

Nevertheless, in most of the computations good results are obtained neglecting this shear lag effect for a PZT sensor [33, 155]. Completing this subsection it is noted that the use of PWASs for measuring Lamb waves is a good choice not only because of their lightness and mobility, but also because the in-plane strains measured by them in general are more sensitive to the fundamental symmetric fundamental wave mode S_0 as out-of-plane displacements sensed by a laser vibrometer (because of higher in-plane magnitude of the symmetric wave mode compared with its out-of-plane magnitude).

2.4.5 Various types and shapes of piezo-actuators

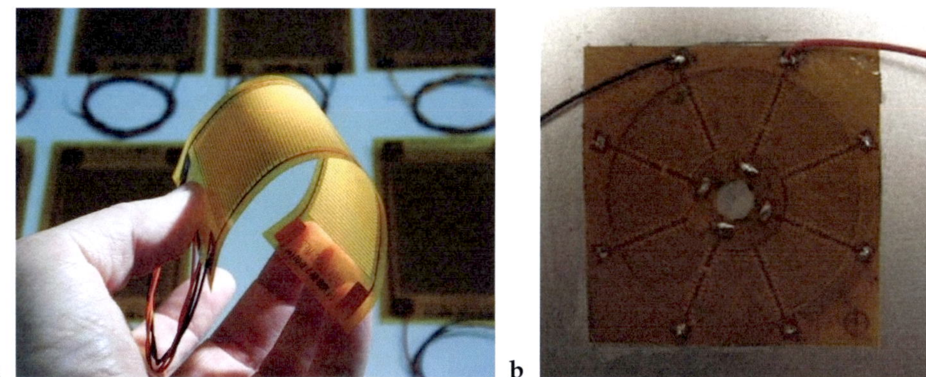

Figure 2.16: MFC transducer (a) [151] and CLoVER transducer (b) [119]

The use of piezoelecric wafers for activating and receiving Lamb waves show a number of problems which should be solved to achieve a high performance of their use for SHM. For example, it is more effective for SHM to activate only one Lamb wave mode in order to prevent a complicated structure of wave fields and to simplify the analysis of measured results. Note that under single PZT actuation both symmetric and antisymmetric Lamb modes are excited simultaneously. Another problem is to achieve directional waves having the highest amplitudes in desired directions. These problems can be solved by properly choosing the excitation frequency, signal bandwidth, the shape of the actuators, their number and location. The frequently used shapes of PZT wafers are briefly presented in this section.

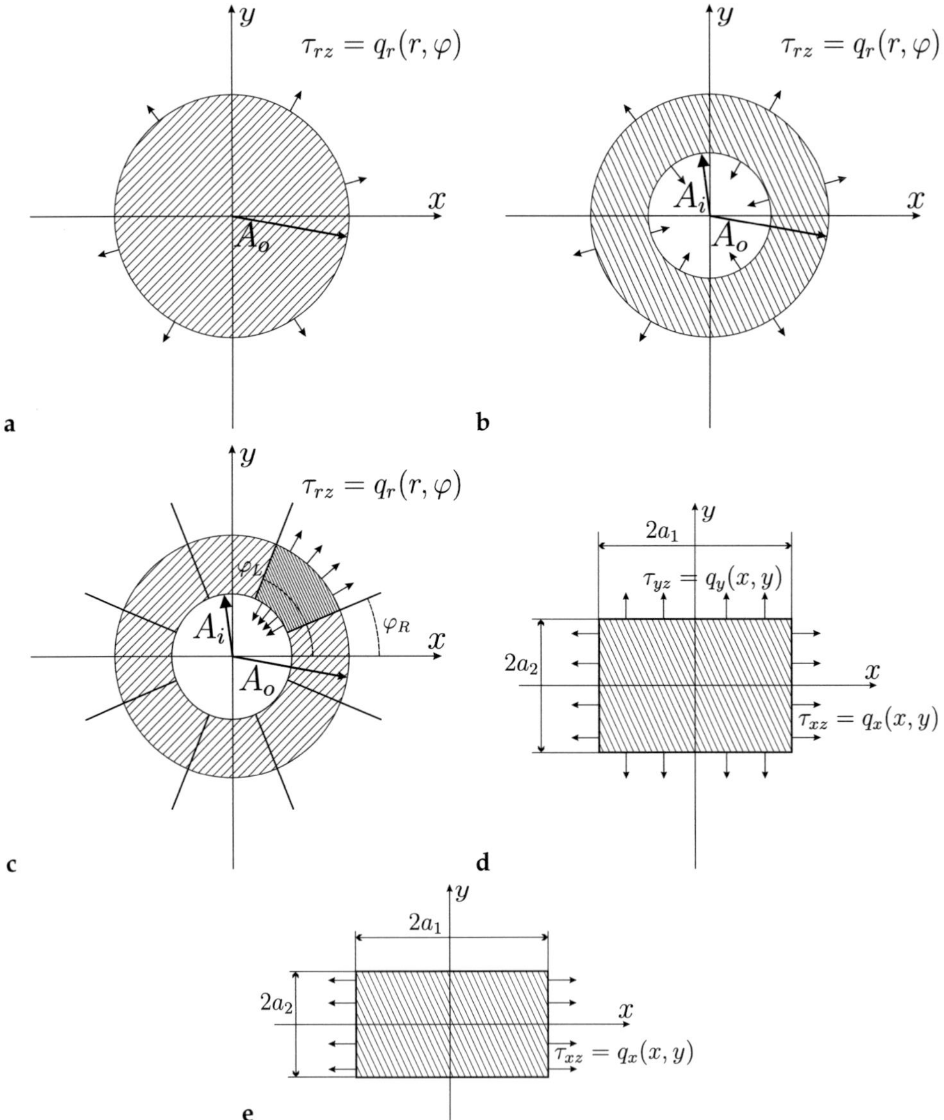

Figure 2.17: Most frequently used shapes of piezoelectric actuators: circular (a), ringshaped (b), CLoVER sector (c), rectangular (d) and MFC (e)

The mostly used PWASs for SHM have a *circular* shape (Figure 2.17a) due to its axis-symmetric properties allowing to simplify the analysis of Lamb waves especially in isotropic plates because the waves generated by PZT wafers of circular shape in isotropic plates are omnidirectional. Such piezoactuators in terms of pin-force model can be well described by the following interface stresses

$$\tau_{xz} = \tau_0 \delta(r - A_o) \cos \varphi, \quad \tau_{yz} = \tau_0 \delta(r - A_o) \sin \varphi, \quad \sigma_z = 0, \text{ for } z = 0, \tag{2.83}$$

where A_o is the radius of the wafer and δ is the delta-function. The model (2.83) assumes the concentration of the radial shear stresses on the outer boundary of the PZT wafer, which are directed outwards to the actuator. Similarly, instead of circular actuators *ring-shaped* PZT wafers are used (Figure 2.17b). These actuators are also axis-symmetric, according to the pin-force model the energy radiated by them is transfered across the inner boundary of the ring in direction of the ring center point and across the outer boundary outwards, that gives

$$\begin{aligned}
\tau_{xz} &= \tau_0 (\delta(r - A_o) - \delta(r - A_i)) \cos \varphi, \\
\tau_{yz} &= \tau_0 (\delta(r - A_o) - \delta(r - A_i)) \sin \varphi, \\
\sigma_z &= 0,
\end{aligned} \tag{2.84}$$

for $z = 0$, where A_o and A_i are the outer and the inner radii of the ring-shaped wafer respectively. The ring-shaped actuators can also be combined with circular actuator placed inside of the ring to the *dual* PZT, where both components are excited independently or simultaneously in order to produce the different wave patterns as it was done for Lamb wave mode decomposition in [155].

Another type of actuators has a rectangular shape. Such actuators are not axis-symmetric and hence produce non-homogeneous wave fields. The classical rectangular piezos can be described by the shear stresses concentrated on their boundaries (Figure 2.17d) as

$$\begin{aligned}
\tau_{xz} &= \tau_0 (\delta(x - a_1) - \delta(x + a_1))(H(y + a_2) - H(y - a_2)), \\
\tau_{yz} &= \tau_0 (H(x + a_1) - H(x - a_1))(\delta(y - a_2) - \delta(y + a_2)), \\
\sigma_z &= 0,
\end{aligned} \tag{2.85}$$

for $z = 0$, where H is the Heaviside step function, $2a_1$ and $2a_2$ are the dimensions of the piezo along x- and y- axes respectively. The wave fields excited by these actuators are much more complicated than wave fields produced by circular and ring-shaped actuators, however in their far-field they are practically identical [33].

The NASA's Langley Research Center developed another type of piezoelecric devices, the Macro-Fiber Composites (MFCs) (Figure 2.16a). MFCs are actuators with unidirectionally aligned fibers, what results in the shear stress only along macro-fibers (Figure 2.17e), so the corresponding pin-force model is given by

$$\tau_{xz} = \tau_0(\delta(x - a_1) - \delta(x + a_1))(H(y + a_2) - H(y - a_2)), \quad \tau_{yz} = 0, \quad \sigma_z = 0, \quad (2.86)$$

for $z = 0$ with the fibers of MFC parallel to x-axis. The Lamb waves caused by an MFC actuator are directed mainly along fiber direction of the MFC patch. Hence, MFC actuators can be used for producing guided waves with strong amplitudes in desired directions. Note that conventional rectangular shaped PWAS with high length-to-width ratio can also generate unidirectional waves [33], producing a weaker signal compared with the MFC and requiring much more energy than an MFC. Another advantage of the MFC is its light weight and flexibility.

An alternative concept for excitation of guided waves in SHM systems is a composite long-range variable-direction emitting radar (*CLoVER*) transducer [119]. It consists of a number of wedge-shaped actuators arranged in a circular array (Figure 2.16b). Geometrically this array has a ring shape, each CLoVER transducer represents one sector of the ring (Figure 2.17c, Figure 2.16b). Each sector in the transducer can independently act as actuator and sensor. The parameters of CLoVER transducers are given by inner A_i and outer A_o radii of the actuators in the array and by the number n_c of partition sectors in the array. The latter parameter allows to define the azimuthal span (in degrees) of CLoVER transducers: $\theta = 360°/n_c$. The pin-force model for one active sector of the CLoVER transducer takes the form [119]

$$\begin{aligned}
\tau_{xz} &= \tau_0(\delta(r - A_i) - \delta(r - A_o))(H(\varphi - \varphi_L) - H(\varphi - \varphi_R))\cos\varphi, \\
\tau_{yz} &= \tau_0(\delta(r - A_i) - \delta(r - A_o))(H(\varphi - \varphi_L) - H(\varphi - \varphi_R))\sin\varphi, \qquad (2.87) \\
\sigma_z &= 0,
\end{aligned}$$

for $z = 0$, where the shear stresses are non-zero only along the wafer's radial edges (see Figure 2.17c) and the active sector is located between the angles φ_L and φ_R: $\varphi_L - \varphi_R = \theta$. The main advantage of the CLoVER transducer is the strong directionality of the wave pattern which is controlled by selecting the desired sector. The pin-force models for PZT wafers of many other shapes are studied in detail in work [130].

Remark 2.2 *The load functions presented in this section for the pin-force model of piezoelectric actuators are given using the definition of delta function for the boundaries of the PZT wafer. If the load function is given only in a single point, it describes a concentrated point load, namely point force - the simplest and very important case of surface loading as it will be clear from the next sections.*

3 Wavenumber-frequency domain solution of the wave propagation problem

3.1 Application of integral approach to the elastodynamic problem

3.1.1 Basics of integral approach

The mathematical boundary value problems considered in the previous chapter for modelling of wave propagation in laminated composites can be solved by such approaches as Ray Tracing Algorithm (used in Lamb wave tomography) [55], FEM [154], SEM [79], Spectral Finite Element Method (SFEM) [44], Semi-Analytical Finite Element Method (SAFEM) [47, 48] and Finite Difference (FD) method [72, 73, 126]. However, all these techniques are time-consuming and do not take into account the wave structure of the problem.

Another method, proven to be an appropriate one is an integral approach for semi-infinite structures like a layer or a plate. In many mathematical problems the Fourier, Hankel, Laplace, Mellin and Radon [147] transforms found an application. In this thesis the application of the Fourier transform to the elastodynamic problem (2.5), (2.10), (2.12) and approximate models using MLPT (2.29) and CLPT (2.36) is considered. The Fourier transform $\mathcal{F}_t$ of the displacement vector $\mathbf{u}(\mathbf{x}, t)$ with respect to time variable t assuming the zero displacement $\mathbf{u}(\mathbf{x}, t)$ and velocity $\dot{\mathbf{u}}(\mathbf{x}, t)$ for $t \leq 0$ (2.14) is defined as

$$\mathbf{u}(\mathbf{x}, \omega) = \mathcal{F}_t[\mathbf{u}(\mathbf{x}, t)] = \int_0^\infty \mathbf{u}(\mathbf{x}, t) \, e^{i\omega t} \, dt. \tag{3.1}$$

The inverse Fourier transform $\mathcal{F}_\omega^{-1}$ of vector of the harmonic steady-state displacements $\mathbf{u}(\mathbf{x}, \omega)$, gives the original transient displacement vector (with only real values) in the form

$$\mathbf{u}(\mathbf{x},t) \;=\; \mathcal{F}_\omega^{-1}[\mathbf{u}(\mathbf{x},\omega)] = \frac{1}{2\pi} \int\limits_{-\infty}^{\infty} \mathbf{u}(\mathbf{x},\omega)\, e^{-i\omega t}\, d\omega \tag{3.2}$$

$$= \;\frac{1}{\pi}\,\mathrm{Re} \int\limits_{0}^{\infty} \mathbf{u}(\mathbf{x},\omega)\, e^{-i\omega t}\, d\omega.$$

Remark 3.1 *With respect to the time variable t, the Laplace transform can be applied analogously (e.g., see [10]).*

Similarly to the Fourier transform with respect to time variable, it is possible to apply the Fourier transform with respect to space variables x and y because both space variables x and y take values from $-\infty$ to $+\infty$, the boundaries of the host structure are parallel to the corresponding xy-plane, the equations of motion are linear and all coefficients in the equations are constant. In the following the Fourier transform with respect to space variables x and y is denoted with capital letters, e.g. $\mathbf{U}(\alpha_1,\alpha_2,z,\omega)$:

$$\mathbf{U}(\alpha_1,\alpha_2,z,\omega) = \mathcal{F}_{x,y}[\mathbf{u}(x,y,z,\omega)] = \int\limits_{-\infty}^{\infty}\int\limits_{-\infty}^{\infty} \mathbf{u}(x,y,z,\omega)\, e^{i(\alpha_1 x + \alpha_2 y)}\, dx dy. \tag{3.3}$$

Note that the application of the Fourier transform to the function of displacements requires that the displacements tend to zero if $x,y \to \infty$, see Equation (2.13).

The parameters of the Fourier domain α_1 and α_2 are called wavenumber variables[1]. The corresponding inverse Fourier transform is given by the double integral

$$\mathbf{u}(x,y,z,\omega) = \mathcal{F}_{\alpha_1,\alpha_2}^{-1}[\mathbf{U}(\alpha_1,\alpha_2,z,\omega)] = \int\limits_{-\infty}^{\infty}\int\limits_{-\infty}^{\infty} \mathbf{U}(\alpha_1,\alpha_2,z,\omega)\, e^{-i(\alpha_1 x + \alpha_2 y)}\, d\alpha_1 d\alpha_2. \tag{3.4}$$

Remark 3.2 *The use of Fourier integral transform requires the integrability of displacements:*

$$\int\limits_{-\infty}^{\infty}\int\limits_{-\infty}^{\infty} |\mathbf{u}(x,y,z,\omega)|\, dx dy < \infty. \tag{3.5}$$

If the derivatives of the given function $\mathbf{u}(x,y,z,t)$ with respect to space x and y or time t variables are integrable functions, i.e. they satisfy (3.5) and their Fourier transform into wavenumber-frequency domain is obtained as follows:

$$\mathcal{F}_{x,y,t}[\partial^{k+m+p}\mathbf{u}(x,y,z,\omega)/\partial x^k \partial y^m \partial t^p] = (-i\alpha_1)^k(-i\alpha_2)^m(-i\omega)^p\mathbf{U}(\alpha_1,\alpha_2,z,\omega). \tag{3.6}$$

---

[1] Some authors use for wavenumber variables the notations k_x and k_y, however hereinafter the notations α_1 and α_2 introduced in [10] are used.

The application of an integral approach consists in an application of the chosen integral transform to the equations of the problem, corresponding boundary conditions, finding the solution of the problem in the transformed domain and performing the inverse transform to the found solution of the problem. These steps for a problem of wave propagation in a laminated composite plate under a surface loading are discussed in detail in this work.

3.1.2 Use of integral approach for obtaining the solution of a elastodynamic problem in a transformed domain

For a homogeneous elastic plate it is known [10] that the displacements caused by a surface load $\mathbf{q}(x,y)v(t)$ with a transient signal of duration $[0,t_0]$ acting in domain Ω can be found using the triple convolution integral [10, 34] as

$$\mathbf{u}(\mathbf{x},t) = \int_0^{t_0} v(\tau) \iint_\Omega \mathbf{k}_t(\mathbf{x} - \boldsymbol{\xi}, t - \tau)\mathbf{q}(\boldsymbol{\xi})\, \mathrm{d}\boldsymbol{\xi}\, \mathrm{d}\tau, \tag{3.7}$$

where $\mathbf{k}_t(\mathbf{x},t)$ is the 3×3 Green's matrix of the problem, where column j ($j = 1,2,3$) of the matrix describes the response of the elastic plate on the action of a point source of the form $q_j = \delta(x)\delta(y)$ with a unit impulse $v(t) = \delta(t)$ as an excitation signal. Application of the Fourier transform with respect to t to both sides of Equation (3.7) and use of the properties of the convolution integral [144] result in

$$\mathbf{u}(\mathbf{x},\omega) = V(\omega) \iint_\Omega \mathbf{k}(\mathbf{x} - \boldsymbol{\xi},\omega)\mathbf{q}(\boldsymbol{\xi})\, \mathrm{d}\boldsymbol{\xi}, \tag{3.8}$$

where $V(\omega)$ is a frequency spectrum of the excitation signal and $\mathbf{k}(\mathbf{x} - \boldsymbol{\xi},\omega)$ is the 3×3 Green's matrix of the corresponding harmonic steady-state problem. The Fourier transform of both sides of Equation (3.8) with respect to space variables x and y yields

$$\mathbf{U}(\alpha_1,\alpha_2,z,\omega) = V(\omega)\mathbf{K}(\alpha_1,\alpha_2,z,\omega)\mathbf{Q}(\alpha_1,\alpha_2) = V(\omega)\mathbf{U}_h(\alpha_1,\alpha_2,z,\omega), \tag{3.9}$$

where $\mathbf{Q}(\alpha_1,\alpha_2)$ describes the wavenumber's domain represention of the surface load vector $\mathbf{q}(\mathbf{x})$, $\mathbf{K}(\alpha_1,\alpha_2,z,\omega)$ is the 3×3 Green's matrix of the problem in the wavenumber-frequency domain and $\mathbf{U}(\alpha_1,\alpha_2,z,\omega)$ represents the solution of the transient problem in the wavenumber-frequency domain. The vector $\mathbf{U}_h(\alpha_1,\alpha_2,z,\omega)$ is also called the *steady-state* harmonic solution of the problem, i.e. when the semi-infinite laminated plate oscillates harmonically with circular frequency ω. In this case the value $\mathbf{u}_h(\mathbf{x},\omega)\,\mathrm{e}^{-\mathrm{i}\omega t}$ corresponds to the wave field obtained under harmonic excitation, where $\mathbf{u}_h(\mathbf{x},\omega) = \mathcal{F}_{\alpha_1,\alpha_2}^{-1}[\mathbf{U}_h(\alpha_1,\alpha_2,z,\omega)]$. For convenience the subscript h is omitted and the solution of the harmonic problem is denoted as

$$\mathbf{U}(\alpha_1,\alpha_2,z,\omega) \equiv \mathbf{U}_h(\alpha_1,\alpha_2,z,\omega).$$

Finally, in the original time-space domain the solution can be obtained as an inverse transform [10]

$$\mathbf{u}(x,y,z,t) \;=\; \frac{1}{4\pi^3}\,\mathrm{Re}\int\limits_0^\infty V(\omega) \tag{3.10}$$

$$\times \iint\limits_{\Gamma_1\Gamma_2} \mathbf{K}(\alpha_1,\alpha_2,z,\omega)\mathbf{Q}(\alpha_1,\alpha_2)\,e^{-i(\omega t+\alpha_1 x+\alpha_2 y)}\,d\alpha_1 d\alpha_2\,d\omega,$$

where Γ_1 and Γ_2 are some contours, mostly coinciding with the real axes with respect to α_1 and α_2, and deviating into the corresponding complex planes while bypassing the real singularities of the components of Green's matrix $\mathbf{K}(\alpha_1,\alpha_2,z,\omega)$. However, it is more convenient to consider the integral (3.10) in polar coordinates (2.43), i.e.

$$\mathbf{u}(r,\varphi,z,t) \;=\; \frac{1}{4\pi^3}\,\mathrm{Re}\int\limits_0^\infty V(\omega) \tag{3.11}$$

$$\times \int\limits_0^{2\pi}\int\limits_{\Gamma^+(\gamma)} \mathbf{K}(\alpha,\gamma,z,\omega)\mathbf{Q}(\alpha,\gamma)\,e^{-i(\omega t+\alpha r\cos(\gamma-\varphi))}\alpha\,d\alpha d\gamma\,d\omega.$$

As it is known for elasticity problems with isotropic properties [10], in case of the pure elastic problem without taking into account the damping (or viscosity) of the host structure, its Green's matrix components have some singularities, *poles*[1,2] on the real axes α_1 and α_2, and the integration directly over the real axes is not possible. The evaluation of integrals (3.11) is performed along some contour $\Gamma^+(\gamma)$ mostly coinciding with a positive real semi-axis while bypassing the real poles from below or from above. Taking different integration contours in (3.11) many different solutions of the problem are possible. In order to obtain a unique solution of the problem using an integral approach, some additional conditions on the solution of the problem are required.

3.1.3 Principle of limiting absorption

The additional conditions on the solution of the problem of pure elasticity are also known as *radiation* conditions or *radiation principles*. There are some principles known including Sommerfeld principle, principle of limiting amplitude, Mandelstamm's principle and *principle of limiting absorption* as it is given in [10]. In this work the latter principle, namely the principle of limiting absorption is used. This principle ensures physical meaning and uniqueness of the given solution. According to this principle,

---

[1]In following the poles of Green's matrix components will also be called *poles of Green's matrix*.
[2]Note that in case of the layered half-space, i.e. $z_{N+1} = -\infty$, additionally the branch points are presented.

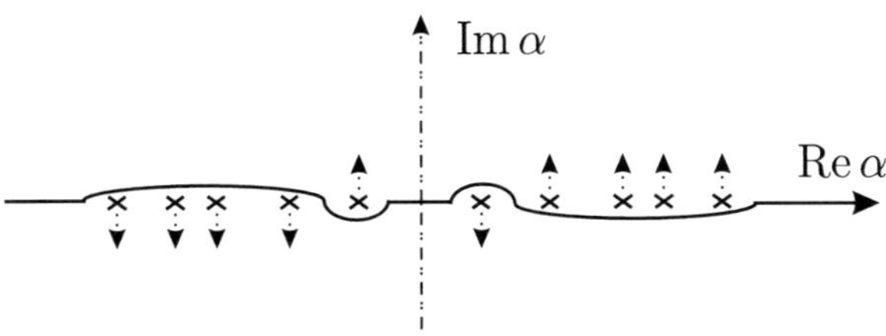

Figure 3.1: Choice of the integration contour $\Gamma^+(\gamma)$ according to the principle of limiting absorption

an infinitesimal damping $\varepsilon > 0$ proportional to the particle velocity as $\varepsilon \partial \mathbf{u}/\partial t$ is introduced into the model and the solution of the related problem is found to be $\mathbf{u}_\varepsilon$. By taking $\varepsilon \to 0$ the displacement vector $\mathbf{u}_\varepsilon(\mathbf{x})$ converges uniformly [144] to the corresponding solution of pure elastic problem

$$\mathbf{u}_\varepsilon(\mathbf{x}) \Rightarrow \mathbf{u}(\mathbf{x}). \tag{3.12}$$

Due to the introduction of infinitesimal damping the positive real poles of Green's functions attain a positive imaginary part, so the integration contours bypass them from below. Such poles are called *regular*. If a positive real pole attains a negative imaginary part, it is bypassed from above. Such poles have anomalous dispersion and are usually called *irregular* poles. The same choice of the integration contour remains also in case of pure elastic solid. The contour $\Gamma^+(\gamma)$ bypassing the real poles in a complex plane according to this principle is shown in Figure 3.1.

Remark 3.3 *The introduction of infinitesimal damping is equivivalent to the introduction of the complex angular frequency in the form* $\omega_\varepsilon^2 = \omega^2 + i\omega\varepsilon/\varrho$ *with a positive imaginary part.*

However, there are some frequencies at which two real poles are equal, i.e. a pole of second order is occuring, and the uniform limit (Equation (3.12)) of the displacement $\mathbf{u}_\varepsilon(\mathbf{x})$ does not exist. In this case according to the principle of limiting absorption [10, 144] the integration contour $\Gamma^+(\gamma)$ goes through this pole and the wavenumber integral in (3.11) has to be understood as Cauchy principal value (PV) [144], i.e.

$$\mathbf{u}(r,\varphi,z,\omega) = PV\frac{1}{4\pi^2} \int\limits_0^{2\pi} \int\limits_{\Gamma^+(\gamma)} \mathbf{K}(\alpha,\gamma,z,\omega)\mathbf{Q}(\alpha,\gamma)\,e^{-i\alpha r\cos(\gamma-\varphi)}\alpha\,d\alpha d\gamma. \tag{3.13}$$

The frequencies, at which the real poles of second order occur, are called *natural* or *resonance*[1] frequencies. Then, the following theorem holds for the displacement fields at the resonance frequencies [10]:

---

[1]Since the pure elastic structure is considered.

Theorem 3.4 *Suppose $k_0(\gamma)$ is a real pole of Green's matrix of second order. If the vector of surface load $\mathbf{Q}(\alpha, \gamma)$ satisfies*

$$\int_0^{2\pi} |\mathbf{Q}(k_0(\gamma), \gamma)| \, d\gamma \neq 0, \tag{3.14}$$

then the displacements at this frequency are growing as $t^{1/2}$ if $t \to \infty$.

Since the resonance frequencies for the regular poles occur only at cut-off frequencies of higher-order wave modes, i.e. $k_0(\gamma) \equiv 0$, for boundedness of the displacements at the resonance frequency it is necessary and sufficient [10] that

$$\mathbf{Q}(0,0) = 0. \tag{3.15}$$

The value $\mathbf{Q}(0,0)$ is also called the main vector of surface stresses. Note that in practice the structures exhibit non-zero damping and the displacements are bounded even at the resonance frequencies. Moreover, in an implementation on PC due to numerical errors the excitation with a resonance frequency is practically impossible.

The solution obtained applying the integral approach in form (3.11) requires algorithms of a quick computation of improper iterated integrals and algorithms of a quick evaluation of surface load vector $\mathbf{Q}(\alpha, \gamma)$ and Green's matrix $\mathbf{K}(\alpha, \gamma, z, \omega)$ in the Fourier domain. These procedures are discussed in detail in this work. The driving signal spectrum $V(\omega)$ can be obtained in most cases in an analytical form (see section 2.4.2).

3.1.4 Fourier transform of the surface load vector-functions

The use of an integral approach for the solution of wave propagation problem needs the representation of the surface load $\mathbf{q}(\mathbf{x})$ in the wavenumber domain. In some cases load functions can be evaluated in the Fourier domain using analytical formulas. The Fourier domain representations of load functions given in section 2.4.5 are described below. These wavenumber transforms can be used to analyse the excitation characteristics of surface sources [130]. Additionally, the way how to compute the wavenumber representation for the surface sources of complicated geometries is presented.

3.1.4.1 Point source

A point force acting along x_k-axis at point $\mathbf{x}_0 = (x_0, y_0, 0) = (r_0 \cos \varphi_0, r_0 \sin \varphi_0, 0)$ is given mathematically by the δ-function. Application of the double Fourier transform results in

$$
\begin{aligned}
Q_k(\alpha_1, \alpha_2) &= \int_{-\infty}^{\infty} \int_{-\infty}^{\infty} \delta(x - x_0)\delta(y - y_0) \, e^{i(\alpha_1 x + \alpha_2 y)} \, dxdy \\
&= e^{i(\alpha_1 x_0 + \alpha_2 y_0)} = e^{i\alpha r_0 \cos(\gamma - \varphi_0)}.
\end{aligned}
\tag{3.16}
$$

3.1.4.2 Circular and ring-shaped PZT wafer

The action of a circular piezo-actuator of radius A_o according to (2.83) in the wavenumber domain can be represented (letting $Q_3 = 0$) in the form

$$Q_1(\alpha, \gamma) = 2\pi i \tau_0 A_o J_1(A_o \alpha) \cos \gamma, \tag{3.17}$$
$$Q_2(\alpha, \gamma) = 2\pi i \tau_0 A_o J_1(A_o \alpha) \sin \gamma,$$

where $J_1(A_o \alpha)$ is the Bessel function of the first kind.

The action of the ring-shaped PZT wafer (2.84) can be assumed to be composed of two circular actuators of different radii and with a different directivity: with a larger radius A_o outwards and with a smaller A_i radius inwards:

$$Q_1(\alpha, \gamma) = 2\pi i \tau_0 [A_o J_1(A_o \alpha) - A_i J_1(A_i \alpha)] \cos \gamma, \tag{3.18}$$
$$Q_2(\alpha, \gamma) = 2\pi i \tau_0 [A_o J_1(A_o \alpha) - A_i J_1(A_i \alpha)] \sin \gamma.$$

3.1.4.3 Rectangular PZT wafers and MFC transducers

A rectangular piezo-actuator of dimensions $2A_1$ and $2A_2$ along the x- and y-axes, respectively modelled as shear stresses concentrated on its boundaries (2.85), is given in the wavenumber domain by $Q_1 = \tau_0 Q / \sin \gamma$, $Q_2 = \tau_0 Q / \cos \gamma$ and $Q_3 = 0$ where

$$Q(\alpha, \gamma) = 4i \sin(A_1 \alpha \cos \gamma) \sin(A_2 \alpha \sin \gamma) / \alpha. \tag{3.19}$$

A similar representation is obtained for an MFC piezo-actuator according to (2.86) aligned along the x-axis:

$$Q_1 = 4i\tau_0 \sin(A_1 \alpha \cos \gamma) \sin(A_2 \alpha \sin \gamma) / (\alpha \sin \gamma), \quad Q_2 = 0, \quad Q_3 = 0. \tag{3.20}$$

3.1.4.4 CLoVER transducer

In contrast to the previous examples, the transform of the load function (2.87) of the CLoVER transducer into the wavenumber domain cannot be performed analytically. However, using Jacobi-Anger expansion (Equation (B.12) in Appendix B.3), the double Fourier transform of (2.87) takes the form

$$Q_1(\alpha, \gamma) = \tau_0 \sum_{n=-\infty}^{\infty} i^n \chi_n(\alpha) \int_{\varphi_R}^{\varphi_L} (H(\varphi - \varphi_R) - H(\varphi - \varphi_L)) \, e^{in(\varphi - \gamma)} \cos \varphi \, d\varphi,$$

$$Q_2(\alpha, \gamma) = \tau_0 \sum_{n=-\infty}^{\infty} i^n \chi_n(\alpha) \int_{\varphi_R}^{\varphi_L} (H(\varphi - \varphi_R) - H(\varphi - \varphi_L)) \, e^{in(\varphi - \gamma)} \sin \varphi \, d\varphi, \tag{3.21}$$

$$Q_3(\alpha, \gamma) = 0,$$

where $\chi_n(\alpha) = (A_o J_n(\alpha A_o) - A_i J_n(\alpha A_i))$, and the active sector with inner radius A_i and outer radius A_o is located between the angles φ_L and φ_R (see Figure 2.17c). The Fourier transform of the load functions results in

$$Q_j = \sum_{n=-\infty}^{\infty} i^n c_n^{(j)} \, e^{-in\gamma} \chi_n(\alpha), \tag{3.22}$$

where the coefficients $c_n^{(j)}$ are expressed as follows:

$$c_{-1}^{(1)} = \frac{1}{2}\left(\varphi_L - \varphi_R + i\frac{e^{-2i\varphi_L} - e^{-2i\varphi_R}}{2}\right), \quad c_1^{(1)} = \frac{1}{2}\left(\varphi_L - \varphi_R - i\frac{e^{2i\varphi_L} - e^{2i\varphi_R}}{2}\right),$$

$$c_n^{(1)} = \frac{1}{2i(n^2-1)} \times \Big[(n-1)\left(e^{i(n+1)\varphi_L}\right. \tag{3.23}$$
$$\left. - e^{i(n+1)\varphi_R}\right) + (n+1)\left(e^{i(n-1)\varphi_L} - e^{i(n-1)\varphi_R}\right)\Big], \quad |n| \neq 1$$

$$c_{-1}^{(2)} = \frac{-i}{2}\left(\varphi_L - \varphi_R - i\frac{e^{-2i\varphi_L} - e^{-2i\varphi_R}}{2}\right),$$

$$c_1^{(2)} = \frac{-i}{2}\left(-\varphi_L + \varphi_R - i\frac{e^{2i\varphi_L} - e^{2i\varphi_R}}{2}\right),$$

$$c_n^{(2)} = -\frac{1}{2(n^2-1)} \times \Big[(n-1)\left(e^{i(n+1)\varphi_L}\right.$$
$$\left. - e^{i(n+1)\varphi_R}\right) - (n+1)\left(e^{i(n-1)\varphi_L} - e^{i(n-1)\varphi_R}\right)\Big], \quad |n| \neq 1.$$

Note that the summation of infinite number of terms in series (3.22) is not possible for numerical computation and the series are truncated by a certain number N_c. The choice of this number N_c depends on the speed of convergence of the series and the accuracy of the Bessel function evaluation, because for large values of $A_o\alpha$ and n the numerical algorithm for the calculation of the Bessel function becomes to be unstable, and the corresponding values should be computed using asymptotic expressions (B.14), given in Apppendix B.3.

Therefore, the series representation (3.22) for a CLoVER sector allows to describe the partial case of the CLoVER transducer - a ring-shaped piezo-actuator, where $\varphi_L = \varphi_R + 2\pi$. In this case the coefficients in (3.23) are obtained as follows:

$$c_n^{(1)} = c_n^{(2)} = 0, \quad |n| \neq 1, \quad c_{-1}^{(1)} = c_1^{(1)} = \pi, \quad c_{-1}^{(2)} = \pi, \quad c_1^{(2)} = -\pi,$$

$$Q_1(\alpha,\gamma) = \tau_0 i^{-1} \pi \, e^{i\gamma} \chi_{-1}(\alpha) + \tau_0 i\pi \, e^{-i\gamma} \chi_1(\alpha) = \tau_0 i\pi \chi_1(\alpha)\left[e^{i\gamma} + e^{-i\gamma}\right]$$
$$= 2\pi i \tau_0 \left(A_o J_1(\alpha A_o) - A_i J_1(\alpha A_i)\right) \cos\gamma, \tag{3.24}$$

$$Q_2(\alpha,\gamma) = \tau_0 i^{-1} \pi \, e^{i\gamma} \chi_{-1}(\alpha) + \tau_0 i(-\pi) \, e^{-i\gamma} \chi_1(\alpha) = -i\pi\tau_0 \chi_1(\alpha) i\left[e^{i\gamma} - e^{-i\gamma}\right]$$
$$= 2\pi i \tau_0 \left(A_o J_1(\alpha A_o) - A_i J_1(\alpha A_i)\right) \sin\gamma.$$

It is evident that the latter representation is equivalent to that obtained previously for a ring-shaped source in (3.18).

3.1.4.5 Wavenumber domain representation for a surface source of general type

Representations obtained in the last subsections for some partial cases of surface sources take into account symmetry of their structure and are very useful for quick computation of the displacement field. If the load function $\mathbf{q}(\mathbf{x})$ does not have a symmetry, it can be approximated by a superposition of the pointwise δ-sources [36, 41]

$$\mathbf{q}(x,y) = \delta_s^2 \sum_{j=1}^{N_q} \mathbf{q}_j \delta(x - x_j)\delta(y - y_j), \tag{3.25}$$

where δ_s is a cubature spacing of domain Ω. A representation for the displacement field in terms of Green's matrix of the problem (3.7) is obtained as a superposition of waves, excited by point sources

$$\mathbf{u}(x,y,z) \approx \delta_s^2 \sum_{j=1}^{N_q} \mathbf{k}(\mathbf{x} - \mathbf{x}_j)\mathbf{q}_j. \tag{3.26}$$

This formula ensures a great importance for practical applications of models studying point sources as wave excitation sources. For example, instead of the circular actuator (2.83) its approximation by N_q pointwise δ sources allocated along its boundary $r = A_o$ on the same distance is considered: $\sigma_{rz} = \delta(r - A_o) \sum_{j=1}^{N_q} \delta(\varphi - \varphi_j)$ and $\varphi_j = 2\pi j/N_q$. Due to an axisymmetric distribution of surface stresses along radiation directions they are set $\mathbf{q}_j = 2\pi/N_q$ for all j.

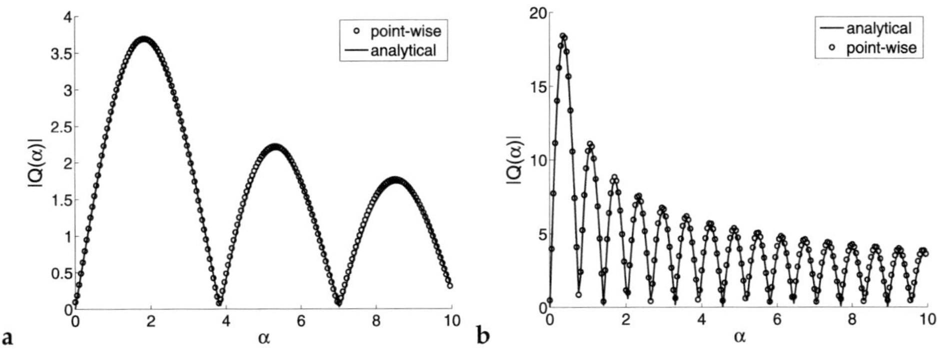

Figure 3.2: Comparison between the analytical ("−") and pointwise ("o") series representations in wavenumber domain (dimensionless form, $\tau_0 = 1$) for two PZT disks with radius $A_o = 1$ (a) and with radius $A_o = 5$ (b)

The representation of this approximation in the wavenumber domain takes the form

$$Q_r(\alpha,\gamma) = \frac{2\pi}{N_q} \sum_{j=1}^{N_q} e^{i\alpha A_o \cos(\gamma - \varphi_j)}. \tag{3.27}$$

The pointwise representation of the surface source (3.27) can be understood as a quadrature rule for a wavenumber representation of surface source [41]. Figure 3.2 illustrates the comparison between the analytical and the pointwise series ($N_q = 64$) representations of a circular piezoactuator in the wavenumber domain for two different radii of actuator $A_o = 1$ and $A_o = 5$. The results are in a well coincidence. However, the accuracy of the series representation should be controlled each time before its use.

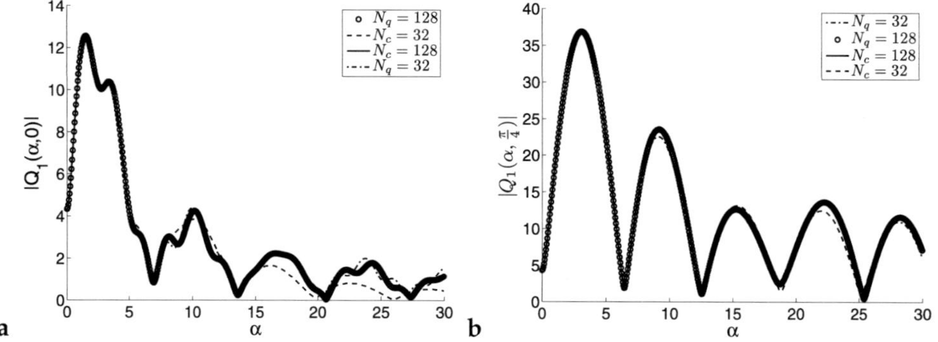

Figure 3.3: Comparison between the series ("$-$" for $N_c = 128$ and "$--$" for $N_c = 32$) and pointwise (("o" for $N_q = 128$ and "$-.-$" for $N_q = 32$) series representations of the $Q_1(\alpha, \gamma)$ (dimensionless form, $\tau_0 = 1$) for a CLoVER transducer of inner radius $A_i = 4$ and outer radius $A_o = 5$ with azimuthal span $\theta = 45°$ and a middle direction aligned along the direction $\varphi = 45°$ calculated in two directions $\gamma = 0°$ (a) and $\gamma = 45°$ (b)

Another example is a CLoVER sector with radial dimensions of $A_i = 4$ and $A_o = 5$, for which the comparison of given expansion with series representation (3.22) and pointwise series representation by setting $\mathbf{q}_j = 2\pi/N_q$ is shown in Figure 3.3. Thereby expansions (3.22), (3.26) are validated. Note that the use of 32 terms in both series representations is not enough for an acceptable accuracy (especially for $\gamma = 0°$) and the use of series representation (3.22) is more sensitive to the choice of insufficient number of terms in corresponding series.

3.2 Green's matrix for a multilayered laminated composite

3.2.1 An algorithm of evaluation of Green's matrix

The formula (3.11) can be used to compute the solution of the wave propagation problem. However, assuming that the frequency spectrum $V(\omega)$ of the excitation signal and the wavenumber representation of the surface load vector $\mathbf{q}(\mathbf{x})$ are known, Green's matrix $\mathbf{K}(\alpha_1, \alpha_2, z, \omega)$ in the wavenumber-frequency domain is required to be computed before the inverse Fourier transform (3.11) can be applied. The algorithm of evaluation of Green's matrix in the Fourier domain for a laminated composite plate with an arbitrary anisotropy of the layers is originally published in *Ultrasonics* Journal [61]. The modification of this algorithm numerically stable for all values of α_1, α_2 and ω containing no growing exponents at any calculation stage is published in *Advances in Theoretical and Applied Mathematics* Journal [58]. This algorithm expands the algorithm suggested originally in [10] for multilayered plates with isotropic layers to the case of triclinic properties of single layers and modifies the algorithm given in [9, 11] for a multilayered anisotropic half-space. Below, the algorithm of evaluation of the frequency-wavenumber domain representation of Green's matrix for an elastodynamic problem (2.5), (2.10), (2.12) is discussed in detail.

At first, the Fourier transforms with respect to time $\mathcal{F}_t$ and space $\mathcal{F}_{xy}$ are applied to the equations of motion of each layer (2.5) and to the boundary (2.10), (2.12) and the interface conditions (2.8). Using the properties of the Fourier transform of derivatives (3.6), the differentiation operators in the Equations (2.5) are replaced by the multipliers

$$\frac{\partial^2}{\partial x^2} \rightarrow (-i\alpha_1)^2, \quad \frac{\partial^2}{\partial x \partial y} \rightarrow (-i\alpha_1)(-i\alpha_2), \quad \frac{\partial^2}{\partial y^2} \rightarrow (-i\alpha_2)^2, \tag{3.28}$$

$$\frac{\partial}{\partial x} \rightarrow -i\alpha_1, \quad \frac{\partial}{\partial y} \rightarrow -i\alpha_2.$$

The partial derivative operator $\partial^2/\partial t^2$ contained in matrix $\mathbf{A}^{(04)}$ (see Appendix A.5, Equation (A.17)) is replaced by the multiplier $(i\omega)^2$. In the following it is assumed that the problem is harmonic steady-state, so that for displacement vector in Fourier domain $\mathbf{U}$ the value $\mathbf{U}_h$ from (3.9) is understood. So, if needed, the solution of the transient problem in the frequency-wavenumber domain is obtained simply by multiplying the harmonic displacement vector $\mathbf{U}(\alpha_1, \alpha_2, z, \omega)$ (i.e. $\mathbf{U}_h(\alpha_1, \alpha_2, z, \omega)$) by the frequency spectrum of the excitation signal $V(\omega)$, see Equation (3.9).

Assuming that the matrix $\mathbf{A}^{(2)}$ (see Appendix A.5, Equation (A.20)) is invertible, Equation (2.5) can be rewritten as

$$\mathbf{U}'' - i[\mathbf{A}^{(2)}]^{(-1)} \left(\mathbf{A}^{(11)}\alpha_1 + \mathbf{A}^{(12)}\alpha_2 \right) \mathbf{U}' \tag{3.29}$$

$$-[\mathbf{A}^{(2)}]^{(-1)} \left(\mathbf{A}^{(01)}\alpha_1^2 + \mathbf{A}^{(02)}\alpha_2^2 + \mathbf{A}^{(03)}\alpha_1\alpha_2 + \mathbf{A}^{(04)} \right) \mathbf{U} = 0.$$

Inroducing the layer number n in the Equation (3.29) and defining the vector $\overline{\mathbf{U}}^{(n)}(\alpha_1, \alpha_2, z, \omega)$ in form

$$\overline{\mathbf{U}}^{(n)} = \left\{ u_1^{(n)}, u_2^{(n)}, u_3^{(n)}, u_1'^{(n)}, u_2'^{(n)}, u_3'^{(n)} \right\}^T, \tag{3.30}$$

where $u_j'^{(n)}$, $j = 1, 2, 3$, is the ordinary derivative of the j-th displacement component with respect to z, a system of ordinary differential equations is obtained

$$\frac{d\overline{\mathbf{U}}^{(n)}}{dz} = \mathbf{B}^{(n)}\overline{\mathbf{U}}^{(n)}, \quad n = 1\ldots N \tag{3.31}$$

with boundary and interface conditions

$$\left.\mathbf{S}^{(1)}\overline{\mathbf{U}}^{(1)}\right|_{z=z_1} = \mathbf{Q}, \tag{3.32}$$

$$\left.\left[\mathbf{R}^{(n)}\overline{\mathbf{U}}^{(n)} - \mathbf{R}^{(n+1)}\overline{\mathbf{U}}^{(n+1)}\right]\right|_{z=z_{n+1}} = \mathbf{0}, \quad n = 1\ldots N-1 \tag{3.33}$$

$$\left.\mathbf{S}^{(N)}\overline{\mathbf{U}}^{(N)}\right|_{z=z_{N+1}} = \mathbf{0}. \tag{3.34}$$

The matrices $\mathbf{B}^{(n)}$, $\mathbf{R}^{(n)}$, $\mathbf{S}^{(n)}$ in Equations (3.31) to (3.34) depend only on the material properties of each layer, the frequency ω and the Fourier variables α_1, α_2. The matrices $\mathbf{B}^{(n)}$ are given by

$$\mathbf{B}^{(n)} = \begin{pmatrix} \mathbf{0} & \mathbf{I} \\ \mathbf{A}^{(n,2)\,-1}\widetilde{\mathbf{A}}_1^{(n)} & i\mathbf{A}^{(n,2)\,-1}\widetilde{\mathbf{A}}_2^{(n)} \end{pmatrix}, \tag{3.35}$$

where

$$\begin{aligned} \widetilde{\mathbf{A}}_1^{(n)} &= \mathbf{A}^{(n,01)}\alpha_1^2 + \mathbf{A}^{(n,02)}\alpha_2^2 + \mathbf{A}^{(n,03)}\alpha_1\alpha_2 - \mathbf{A}^{(n,04)}, \\ \widetilde{\mathbf{A}}_2^{(n)} &= [\mathbf{A}^{(n,11)}\alpha_1 + \mathbf{A}^{(n,12)}\alpha_2]. \end{aligned} \tag{3.36}$$

The matrices $\mathbf{0}$, $\mathbf{I}$ are the 3×3 null and the identity matrix, respectively. The matrices $\mathbf{R}^{(n)}$ characterizing the layer interfaces are given by

$$\mathbf{R}^{(n)} = \begin{pmatrix} \mathbf{I} & \mathbf{0} \\ -i\mathbf{R}_0^{(n)} & \mathbf{A}^{(n,2)} \end{pmatrix} \tag{3.37}$$

where

$$\mathbf{R}_0^{(n)} = \begin{pmatrix} \alpha_1 C_{51}^{(n)} + \alpha_2 C_{56}^{(n)} & \alpha_2 C_{52}^{(n)} + \alpha_1 C_{56}^{(n)} & \alpha_2 C_{54}^{(n)} + \alpha_1 C_{55}^{(n)} \\ \alpha_1 C_{41}^{(n)} + \alpha_2 C_{46}^{(n)} & \alpha_2 C_{42}^{(n)} + \alpha_1 C_{46}^{(n)} & \alpha_2 C_{44}^{(n)} + \alpha_1 C_{45}^{(n)} \\ \alpha_1 C_{31}^{(n)} + \alpha_2 C_{36}^{(n)} & \alpha_2 C_{32}^{(n)} + \alpha_1 C_{36}^{(n)} & \alpha_2 C_{34}^{(n)} + \alpha_1 C_{35}^{(n)} \end{pmatrix}. \tag{3.38}$$

The boundary conditions above and below for the composite with a stress-free bottom boundary[1] (3.34) are represented by the matrices $\mathbf{S}^{(n)}$, $n = 1$, $n = N$

$$\mathbf{S}^{(n)} = \begin{pmatrix} -i\mathbf{R}_0^{(n)} & \mathbf{A}^{(n,2)} \end{pmatrix}. \tag{3.39}$$

---

[1] By means of this algorithm it is also possible to consider a multilayered plate with a fixed bottom boundary or a multilayered half-space.

Thus, the problem considered results in a set of N systems of ODEs with constant coefficients (for fixed Fourier parameters α_1, α_2 and ω) given by the 6×6 matrices $\mathbf{B}^{(n)}$ (3.31) for each layer and the $6 \times N$ system of boundary (3.32), (3.34) and transition conditions (3.33).

Since the problem is linear, it is possible to decompose the Fourier transform of the vector of displacement components with respect to the components of the applied load vector $\mathbf{Q}(\alpha_1, \alpha_2) = \{Q_1, Q_2, Q_3\}$ as follows:

$$\overline{\mathbf{U}}^{(n)}(\alpha_1, \alpha_2, z, \omega) = \overline{\mathbf{U}}_1^{(n)} Q_1 + \overline{\mathbf{U}}_2^{(n)} Q_2 + \overline{\mathbf{U}}_3^{(n)} Q_3. \tag{3.40}$$

In order to find each $\overline{\mathbf{U}}_j^{(n)}$ in the decomposition (3.40) it is necessary to replace the ternary vector $\mathbf{Q}$ in condition (3.32) with the vector $\mathbf{e}_p$, in which 1 is on the j-th position and all other components are equal to zero:

$$\mathbf{S}^{(1)} \overline{\mathbf{U}}_j^{(1)} \Big|_{z=z_1} = \mathbf{e}_j. \tag{3.41}$$

This decomposition (3.40) provides an opportunity to study the characteristics of the laminated composite separately from the influence of the load. The solution of Equation (3.31) leads to the eigenvalue problem

$$(\mathbf{B}^{(n)} - \lambda^{(n)}\mathbf{I})\mathbf{m}^{(n)} = \mathbf{0}. \tag{3.42}$$

In order to investigate the properties of the eigenvalues of $\mathbf{B}^{(n)}$, the matrix $\widetilde{\mathbf{B}}$ with real components for real α_1 and α_2 is introduced as[1]

$$\widetilde{\mathbf{B}} = \begin{pmatrix} \mathbf{0} & \mathbf{I} \\ -\mathbf{A}^{(2)^{-1}}\widetilde{\mathbf{A}}_1 & \mathbf{A}^{(2)^{-1}}\widetilde{\mathbf{A}}_2 \end{pmatrix}. \tag{3.43}$$

Then, (3.42) is rewritten in the form

$$\begin{aligned}
(\mathbf{B} - \lambda\mathbf{I})\mathbf{m} &= \left(\begin{pmatrix} \mathbf{I} & \mathbf{0} \\ \mathbf{0} & i\cdot\mathbf{I} \end{pmatrix} \widetilde{\mathbf{B}} \begin{pmatrix} i\cdot\mathbf{I} & \mathbf{0} \\ \mathbf{0} & \mathbf{I} \end{pmatrix} - \lambda\mathbf{I} \right)\mathbf{m} \\
&= \begin{pmatrix} \mathbf{I} & \mathbf{0} \\ \mathbf{0} & i\cdot\mathbf{I} \end{pmatrix} \left(\widetilde{\mathbf{B}} - (-i)\,\lambda\mathbf{I} \right) \begin{pmatrix} i\cdot\mathbf{I} & \mathbf{0} \\ \mathbf{0} & \mathbf{I} \end{pmatrix} \mathbf{m} \\
&= \begin{pmatrix} \mathbf{I} & \mathbf{0} \\ \mathbf{0} & i\cdot\mathbf{I} \end{pmatrix} \left(\widetilde{\mathbf{B}} - \widetilde{\lambda}\mathbf{I} \right) \widetilde{\mathbf{m}} = \left(\widetilde{\mathbf{B}} - \widetilde{\lambda}\mathbf{I} \right) \widetilde{\mathbf{m}} = \mathbf{0},
\end{aligned} \tag{3.44}$$

where

$$\lambda = i\widetilde{\lambda}, \quad \mathbf{m} = \begin{pmatrix} -i\cdot\mathbf{I} & \mathbf{0} \\ \mathbf{0} & \mathbf{I} \end{pmatrix} \widetilde{\mathbf{m}}. \tag{3.45}$$

---

[1]The upper index n denoting the number of the layer is omitted for convenience

It yields the modified equation of the eigenvalue problem with a real-valued matrix $\widetilde{\mathbf{B}}$ in the form

$$\left(\widetilde{\mathbf{B}} - \widetilde{\lambda}\mathbf{I}\right)\widetilde{\mathbf{m}} = \mathbf{0}. \tag{3.46}$$

The corresponding characteristic polynomial with real coefficients is obtained as

$$\det\left(\widetilde{\mathbf{B}} - \widetilde{\lambda}\mathbf{I}\right) = \det\left(\widetilde{\mathbf{A}}_1 - \widetilde{\lambda}\widetilde{\mathbf{A}}_2 + \widetilde{\lambda}^2\mathbf{A}^{(2)}\right) / \det\left(\mathbf{A}^{(2)}\right) = \sum_{j=0}^{6} a_j\widetilde{\lambda}^j. \tag{3.47}$$

The polynomial (3.47) of degree 6 with real coefficients can have even number of real and complex conjugated roots only. According to (3.45) it yields the imaginary roots and roots with the same imaginary part and with real parts of opposite sign. This means, there are three roots satisfying the conditions

$$\text{Re}\,\lambda_k > 0, \quad \text{or} \quad \text{Im}\,\lambda_k < 0 \text{ if } \text{Re}\,\lambda_k = 0, \quad k = 1, 2, 3. \tag{3.48}$$

The remaining three roots obviously satisfy

$$\text{Re}\,\lambda_k < 0, \quad \text{or} \quad \text{Im}\,\lambda_k > 0 \text{ if } \text{Re}\,\lambda_k = 0, \quad k = 4, 5, 6. \tag{3.49}$$

The last two conditions are also called *Sommerfeld conditions* [135].

Remark 3.5 *For orthotropic layers the coefficients of odd degrees in the polynomial (3.47) are equal to zero: $a_1 = a_3 = a_5 = 0$. It results in the following relation between the roots*

$$\lambda_k = -\lambda_{k+3}, \quad k = 1, 2, 3. \tag{3.50}$$

Remark 3.6 *For isotropic layers two of the roots λ_k are of second degree [10], i.e.*

$$\lambda_1 = -\lambda_4, \quad \lambda_2 = \lambda_3 = -\lambda_5 = -\lambda_6. \tag{3.51}$$

Remark 3.7 *Sommerfeld conditions (3.48), (3.49) lead to the presence of the branch points in case, when instead of the plate a layered half-space is considered [10, 135].*

In case of arbitrary anisotropy of the layer six eigenvalues $\lambda_k^{(n)}$ and six eigenvectors $\mathbf{m}_k^{(n)}$ in (3.42) can be found only using numerical algorithms, in case of transversal-isotropy using relatively complicated analytical formulas [109] and in the isotropic case analytically using simple expressions given in [10]. Note that the computational time needed for the numerical evaluation of the eigenvalues and eigenvectors gives a sufficient contribution to the whole time needed for the computation of the solution in the wavenumber domain.

Using eigenvalues $\lambda_k^{(n)}$ and eigenvectors $\mathbf{m}_k^{(n)}$, the general solution of the ODE (3.31) with boundary conditions at the upper surface of the multilayered plate in form (3.41) given by a unit vector $\mathbf{e}_j$ can be written as

$$\overline{\mathbf{U}}_j^{(n)}(\alpha_1, \alpha_2, z) = \mathbf{M}^{(n)}(\alpha_1, \alpha_2)\widetilde{\mathbf{E}}^{(n)}(\alpha_1, \alpha_2, z)\widetilde{\mathbf{t}}_j^{(n)}, \tag{3.52}$$

where

$$\mathbf{M}^{(n)} = (\mathbf{m}_1^{(n)} \ldots \mathbf{m}_6^{(n)}) \tag{3.53}$$

is a matrix composed from eigenvectors $\mathbf{m}_k^{(n)}$ of the matrix $\mathbf{B}^{(n)}$, the matrix $\widetilde{\mathbf{E}}^{(n)}$ with

$$\widetilde{E}_{kk}^{(n)}(z) = \exp\left(\lambda_k^{(n)} z\right), \quad k = 1 \ldots 6, \tag{3.54}$$

which is a diagonal matrix and $\widetilde{\mathbf{t}}_j^{(n)}$ is a vector of unknown coefficients of the n-th layer.

For $z < 0$ and $\alpha \to \infty$ the eigenvalues are growing

$$\lambda_k^{(n)} \sim C_k^{(n)} |\alpha|, \quad \alpha \to \infty, \tag{3.55}$$

where the $C_k^{(n)}$ are some complex constants, and according to the property $\exp\left(\lambda_k^{(n)} z\right) \to \infty, k > 3$ if $\alpha \to \infty$. In order to avoid growing exponents in (3.54) the relations (3.48), (3.49) are used and instead of the matrix $\widetilde{\mathbf{E}}^{(n)}$ the matrix $\mathbf{E}^{(n)}(z)$ for each z: $z_{n+1} \leq z \leq z_n$ is defined as follows

$$E_{kk}^{(n)}(z) = \exp\left(\lambda_k^{(n)}(z - z_n)\right), \quad k = 1, 2, 3, \tag{3.56}$$

$$E_{kk}^{(n)}(z) = \exp\left(\lambda_k^{(n)}(z - z_{n+1})\right), \quad k = 4, 5, 6.$$

It leads to other unknown coefficients $\mathbf{t}_j^{(n)}$ instead of $\widetilde{\mathbf{t}}_j^{(n)}$, i.e. the displacement vector - instead of (3.52) is evaluated as

$$\overline{\mathbf{U}}_j^{(n)}(\alpha_1, \alpha_2, z) = \mathbf{M}^{(n)}(\alpha_1, \alpha_2)\mathbf{E}^{(n)}(\alpha_1, \alpha_2, z)\mathbf{t}_j^{(n)}. \tag{3.57}$$

Remark 3.8 *If instead of the layered plate a layered half-space is considered, under some certain conditions on λ_k, $k = 1, \ldots, 6$, the terms $k = 4, 5, 6$ in (3.56) can be neglected since they are non-physical. However, the Sommerfeld conditions (3.48), (3.49) are not valid for layers with low symmetries of stiffness properties (as it is stated in [135]) and the principle of limiting absorption is needed to be applied to distinguish the pure imaginary eigenvalues λ_k between two groups of eigenvalues similarly to (3.48) and (3.49).*

Taking into consideration the required type of solution (3.57), on the basis of the boundary conditions (3.33), (3.34), (3.41) a system for defining the unknown coefficients is obtained

$$\mathbf{W}\mathbf{t}_j = \mathbf{e}_j, \quad j = 1,2,3, \tag{3.58}$$

where $\mathbf{t}_j$ is a $6N$ vector consisting of unknown coefficients for all layers, i.e. consisting of all vectors $\mathbf{t}_j^{(n)}$: $\mathbf{t}_j = (\mathbf{t}_j^{(1)^T}, \mathbf{t}_j^{(2)^T} \ldots \mathbf{t}_j^{(N)^T})^T$ and in turn $\mathbf{t}_j^{(n)} = (t_{j,1}^{(n)} \ldots t_{j,6}^{(n)})^T$, $n = 1 \ldots N$. The matrix $\mathbf{W}$ (3.58) does not depend on j and has the form

$$\mathbf{W} = \begin{pmatrix} \mathbf{L}_1^{(1)} & 0 & 0 & \ldots & 0 & 0 \\ \mathbf{G}_2^{(1)} & -\mathbf{G}_2^{(2)} & 0 & \ldots & 0 & 0 \\ 0 & \mathbf{G}_3^{(2)} & -\mathbf{G}_3^{(3)} & \ldots & 0 & 0 \\ \ldots & \ldots & \ldots & \ldots & \ldots & \ldots \\ 0 & 0 & 0 & 0 & \mathbf{G}_N^{(N-1)} & -\mathbf{G}_N^{(N)} \\ 0 & 0 & 0 & 0 & 0 & \mathbf{L}_{N+1}^{(N)} \end{pmatrix}, \tag{3.59}$$

where

$$\begin{aligned} \mathbf{L}_k^{(n)} &= \mathbf{S}^{(n)}\mathbf{M}^{(n)}\mathbf{E}^{(n)}(z_k), \\ \mathbf{G}_k^{(n)} &= \mathbf{R}^{(n)}\mathbf{M}^{(n)}\mathbf{E}^{(n)}(z_k) \end{aligned} \tag{3.60}$$

for $n = 1 \ldots N$, $k = 1 \ldots N+1$, while the matrices $\mathbf{R}^{(n)}$ and $\mathbf{S}^{(n)}$ are given by (3.37) and (3.39). The number of unknown constants as well as the number of equations in the system (3.58) is $6N$. The solution of this system of the linear equations can be found for each j in a recurrent form as

$$\begin{aligned} \mathbf{t}_j^{(N)} &= \begin{pmatrix} \mathbf{L}_1^{(1)}\widetilde{\mathbf{G}} \\ \mathbf{L}_{N+1}^{(N)} \end{pmatrix}^{-1} \begin{pmatrix} \mathbf{e}_j \\ 0 \end{pmatrix}, \quad j = 1,2,3, \\ \widetilde{\mathbf{G}} &= \left(\mathbf{G}_2^{(1)}\right)^{-1}\mathbf{G}_2^{(2)}\left(\mathbf{G}_3^{(2)}\right)^{-1}\mathbf{G}_3^{(3)} \ldots \left(\mathbf{G}_N^{(N-1)}\right)^{-1}\mathbf{G}_N^{(N)}, \\ \mathbf{t}_j^{(n-1)} &= \left(\mathbf{G}_n^{(n-1)}\right)^{-1}\mathbf{G}_n^{(n)}\mathbf{t}_j^{(n)}, \quad n = N, N-1, \ldots 3, 2. \end{aligned} \tag{3.61}$$

The use of the recurrent form of the solution makes this low-cost algorithm similar to the transfer matrix method [52, 137]. Note that the transfer matrix method and this recurrent algorithm become unstable for large values of ω and α since $\mathbf{E}^{(n)}(z) \to 0$ ($z \neq 0$) as $\alpha \to \infty$ and consequently the matrices $\mathbf{G}_k^{(j)}$ tend to be singular [82]. If the system (3.58) will be solved directly without using the sparsity of the matrix $\mathbf{W}$, this algorithm will be similar to a global matrix method [65, 94], which is more time-expensive, however, is stable for large values of ω and α.

Having solved the system (3.58) for each $j = 1, 2, 3$, the functions $\overline{U}_j^{(n)}(\alpha_1, \alpha_2, z, \omega)$ can be found. They allow to rewrite the solution of the problem $\mathbf{U}^{(n)}$ for each layer n in the form

$$\mathbf{U}^{(n)}(\alpha_1, \alpha_2, z, \omega) = \begin{pmatrix} \overline{U}_{11}^{(n)} & \overline{U}_{21}^{(n)} & \overline{U}_{31}^{(n)}, \\ \overline{U}_{12}^{(n)} & \overline{U}_{22}^{(n)} & \overline{U}_{32}^{(n)}, \\ \overline{U}_{13}^{(n)} & \overline{U}_{23}^{(n)} & \overline{U}_{33}^{(n)}, \end{pmatrix} \mathbf{Q}(\alpha_1, \alpha_2) \tag{3.62}$$

$$= \mathbf{K}^{(n)}(\alpha_1, \alpha_2, z, \omega) \mathbf{Q}(\alpha_1, \alpha_2),$$

where in $\overline{U}_{jk}^{(n)}$ the first index j corresponds to the choice of the boundary condition in (3.41) and the second index k describes the number of component of the vector $\mathbf{U}_j^{(n)}$. The matrix $\mathbf{K}^{(n)}(\alpha_1, \alpha_2, z, \omega)$ is usually called Green's matrix (in the wavenumber domain) of the n-th layer of the multilayered composite plate under the excitation at the top surface. It can be used to compute Green's matrix of the whole plate $\mathbf{K}(\alpha_1, \alpha_2, z, \omega)$ in the wavenumber domain in a form given by the formula (3.11) as

$$\mathbf{K}(\alpha_1, \alpha_2, z, \omega) = \left\{ \mathbf{K}^{(n)}(\alpha_1, \alpha_2, z, \omega), \quad z_{n+1} \leq z \leq z_n, \quad n = 1, \dots, N \right\}. \tag{3.63}$$

The columns of the matrix $\mathbf{K}(\alpha_1, \alpha_2, z, \omega)$ in (3.63) are the vectors of displacements excited by concentrated surface loads $\delta(x)\delta(y) \cdot \mathbf{e}_j, j = 1, 2, 3$ along the coordinate axes ($\mathbf{e}_j$ are the corresponding unit vectors).

Remark 3.9 *Besides the displacement vector obtained using an approach described here, according to (3.30) its derivative with respect to the variable z is computed as follows:*

$$\frac{d\mathbf{U}^{(n)}(\alpha_1, \alpha_2, z, \omega)}{dz} = \begin{pmatrix} \overline{U}_{14}^{(n)} & \overline{U}_{24}^{(n)} & \overline{U}_{34}^{(n)} \\ \overline{U}_{15}^{(n)} & \overline{U}_{25}^{(n)} & \overline{U}_{35}^{(n)} \\ \overline{U}_{16}^{(n)} & \overline{U}_{26}^{(n)} & \overline{U}_{36}^{(n)} \end{pmatrix} \mathbf{Q}(\alpha_1, \alpha_2). \tag{3.64}$$

Note that the derivatives of the displacement vector with respect to x, y and t variables can be obtained in the Fourier domain using (3.6).

3.2.2 Green's matrix for a single-layered isotropic plate

The algorithm of Green's matrix computation (section 3.2.1) provides a numerical solution only. A reasonable analytical solution can be obtained only in the case of a single isotropic layer (already in case of two layers it would be not suitable for the implementation on PC). Within the notations adopted by Babeshko [10], Green's matrix for an isotropic layer can be expressed as

$$\mathbf{K}(\alpha_1, \alpha_2, z, \omega) = \begin{pmatrix} -i\left(\alpha_1^2 M + \alpha_2^2 N\right)/\alpha^2 & -i\alpha_1\alpha_2\left(M - N\right)/\alpha^2 & -i\alpha_1 P \\ -i\alpha_1\alpha_2\left(M - N\right)/\alpha^2 & -i\left(\alpha_2^2 M + \alpha_1^2 N\right)/\alpha^2 & -i\alpha_2 P \\ \alpha_1 S/\alpha^2 & \alpha_2 S/\alpha^2 & R \end{pmatrix}. \tag{3.65}$$

The explicit expressions describing the evaluation of the analytical functions $M(\alpha, z, \omega)$, $N(\alpha, z, \omega)$, $P(\alpha, z, \omega)$, $S(\alpha, z, \omega)$ and $R(\alpha, z, \omega)$ may be found in Appendix A.8.

3.2.3 Main properties of Green's matrix

Before starting with the algorithms for evaluation of the improper iterated integrals of the inverse Fourier transform (3.11), which are given in chapter 5, the properties of Green's matrix need to be investigated.

- The first and most important property of Green's matrix components is the presence of the singularities on the real axes, which require the choice of the integration contour with respect to the principle of limiting absorption or other radiation principles. As it follows from the system of linear equations (3.58) by applying Cramer's rule, each component of Green's matrix contains as denominator the determinant of $\mathbf{W}(\alpha_1, \alpha_2, \omega)$:

$$K_{ij}(\alpha_1, \alpha_2, z, \omega) = \tilde{K}_{ij}(\alpha_1, \alpha_2, z, \omega) / \det \mathbf{W}(\alpha_1, \alpha_2, \omega), \quad i, j = 1, 2, 3. \tag{3.66}$$

  Herewith the functions $\tilde{K}_{ij}(\alpha_1, \alpha_2, z, \omega)$ do not have any singularities at finite points. In turn, the function $\det \mathbf{W}(\alpha_1, \alpha_2, \omega)$ contains all finite singularities of Green's matrix as its roots, i.e.

$$\det \mathbf{W}(\alpha_1, \alpha_2, \omega) = 0. \tag{3.67}$$

  This equation is also known as dispersion equation (see section 2.3.2), and allows to describe the properties of the multilayered plate, taking into account the material properties of each layer, the orientations and the thicknesses of the individual layers as well as their stacking sequence as it is studied in detail in section 4.

- The components K_{11}, K_{12}, K_{21}, K_{22} and K_{33} considered as functions of α are symmetric for each γ, all other components K_{13}, K_{23}, K_{31} and K_{32} are antisymmetric, i.e.

$$\mathbf{K}(-\alpha) = \begin{pmatrix} +K_{11}(\alpha) & +K_{12}(\alpha) & -K_{13}(\alpha) \\ +K_{21}(\alpha) & +K_{22}(\alpha) & -K_{23}(\alpha) \\ -K_{31}(\alpha) & -K_{32}(\alpha) & +K_{33}(\alpha) \end{pmatrix}. \tag{3.68}$$

  Note that similar properties with respect to γ can be obtained if γ is replaced by $\gamma + \pi$ since

$$\alpha_1 = \alpha \cos(\gamma + \pi) = -\alpha \cos \gamma, \quad \alpha_2 = \alpha \sin(\gamma + \pi) = -\alpha \sin \gamma. \tag{3.69}$$

- The expression of the solution in form (3.57) can be used to analyse the asymptotic properties of Green's matrix components, which are found to be decaying functions of α

$$K_{ij}(\alpha, \gamma, z = 0, \omega) = \frac{T_{ij}(\gamma, \omega)}{\alpha} + o\left(\alpha^{-1}\right) \sim \frac{T_{ij}(\gamma, \omega)}{\alpha}, \quad i, j = 1, 2, 3. \tag{3.70}$$

- It is worth mentioning that Green's matrix has another important property, which however, is valid only for isotropic laminates: Green's matrix can be decoupled as

$$
\begin{pmatrix}
-\mathrm{i}\left(\alpha_1^2 M + \alpha_2^2 N\right)/\alpha^2 & -\mathrm{i}\alpha_1\alpha_2\left(M - N\right)/\alpha^2 & -\mathrm{i}\alpha_1 P \\
-\mathrm{i}\alpha_1\alpha_2\left(M - N\right)/\alpha^2 & -\mathrm{i}\left(\alpha_2^2 M + \alpha_1^2 N\right)/\alpha^2 & -\mathrm{i}\alpha_2 P \\
\alpha_1 S/\alpha^2 & \alpha_2 S/\alpha^2 & R
\end{pmatrix}
$$
$$
= \begin{pmatrix}
\cos\gamma & \sin\gamma & 0 \\
-\sin\gamma & \cos\gamma & 0 \\
0 & 0 & 1
\end{pmatrix}^{-1}
\mathbf{K}_{\mathrm{ISO}}(\alpha, z, \omega)
\begin{pmatrix}
\cos\gamma & \sin\gamma & 0 \\
-\sin\gamma & \cos\gamma & 0 \\
0 & 0 & 1
\end{pmatrix},
\tag{3.71}
$$

where

$$
\mathbf{K}_{\mathrm{ISO}}(\alpha, z, \omega) = \begin{pmatrix}
-\mathrm{i}M & 0 & -\mathrm{i}\alpha P \\
0 & -\mathrm{i}N & 0 \\
\dfrac{\mathrm{i}S}{\alpha} & 0 & R
\end{pmatrix}.
\tag{3.72}
$$

3.3 Green's matrices for MLPT and CLPT in transformed domain

The algorithm previously described in section 3.2.1 can be used to evaluate Green's matrix in the wavenumber-frequency domain for the elastodynamic wave propagation problem in a laminated composite plate, Equations (2.5), (2.10), (2.12). Similarly, the problems of wave propagation described by Classical and Mindlin Laminated Plate Theories can be solved using an integral approach. The integral transform $\mathcal{F}_{x,y,t}$ of the equation of motion (2.29) of MLPT with respect to variables x, y and t yields

$$
\mathbf{T}_M \mathbf{U}_M = \mathbf{F}_M,
\tag{3.73}
$$

where the partial derivatives with respect to x, y and t in matrices $\mathbf{T}_M$ (Appendix 2.2.3) are replaced by the multipliers according to formula (3.6). Note that the matrix $\mathbf{T}_M$ is not-symmetric, but positive definite Hermitian. The transformed displacement vector $\mathbf{U}_M$ and load vector $\mathbf{F}_M$ are

$$
\begin{aligned}
\mathbf{U}_M(\alpha_1, \alpha_2, \omega) &= \left(U_0, V_0, W_0, \Psi_x, \Psi_y\right)^T, \\
\mathbf{F}_M(\alpha_1, \alpha_2, \omega) &= V(\omega) \\
&\times \left(Q_1(\alpha_1, \alpha_2), Q_2(\alpha_1, \alpha_2), Q_3(\alpha_1, \alpha_2), \frac{h}{2}Q_1(\alpha_1, \alpha_2), \frac{h}{2}Q_2(\alpha_1, \alpha_2)\right)^T.
\end{aligned}
$$

The vector of displacements at mid-plane in the wavenumber-frequency domain $\mathbf{U}_M$ is obtained by simple inversion of the matrix $\mathbf{T}_M$:

$$
\mathbf{U}_M = \mathbf{T}_M^{-1}\mathbf{F}_M.
\tag{3.74}
$$

However, the representations are not of the same form as the solution of the elasto-dynamic problem (3.9). Applying the Fourier transform to the expressions (2.15), the displacements are obtained as

$$
\begin{aligned}
U_1(\alpha_1,\alpha_2,z,\omega) &= U_0(\alpha_1,\alpha_2,\omega) + z\Psi_x(\alpha_1,\alpha_2,\omega), \\
U_2(\alpha_1,\alpha_2,z,\omega) &= V_0(\alpha_1,\alpha_2,\omega) + z\Psi_y(\alpha_1,\alpha_2,\omega), \\
U_3(\alpha_1,\alpha_2,z,\omega) &= W_0(\alpha_1,\alpha_2,\omega),
\end{aligned}
$$

i.e. Green's matrix of the Mindlin Laminated Plate takes the form

$$
\begin{aligned}
\mathbf{U}_{\mathrm{MLPT}}(\alpha_1,\alpha_2,z,\omega) &= \begin{pmatrix} U_{0,1}+z\Psi_{x,1} & U_{0,2}+z\Psi_{x,2} & U_{0,3}+z\Psi_{x,3}, \\ V_{0,1}+z\Psi_{y,1} & V_{0,2}+z\Psi_{y,2} & V_{0,3}+z\Psi_{y,3}, \\ W_{0,1} & W_{0,2} & W_{0,3} \end{pmatrix} \mathbf{Q}(\alpha_1,\alpha_2) \\
&= \mathbf{K}_{\mathrm{MLPT}}(\alpha_1,\alpha_2,z,\omega)\mathbf{Q}(\alpha_1,\alpha_2),
\end{aligned}
\tag{3.75}
$$

where the index j in $U_{0,j}$, $V_{0,j}$, $W_{0,j}$, $\Psi_{x,j}$ and $\Psi_{y,j}$ describes the displacement response of the structure at the mid-plane due to the action of the concentrated shear traction along the x-axis, i.e. $Q_j = 1$, $Q_k = 0$ for $k \neq j$. The components K_{11} and K_{33} of Green's matrix $\mathbf{K}_{\mathrm{MLPT}}$ are compared with those of Green's matrix of the elastodynamic problem in Figure 3.4 for the $[0/90]_s$ laminated composite plate on the basis of CFRP-T700GC/M21 (material properties are given in Table A.1 in Appendix A.9) for the frequency-thickness $f \cdot h = 300\ \mathrm{KHz} \cdot \mathrm{mm}$ and for $\gamma = 45°$ in dependence on α. Here the real singularities of both matrices are located close to each other for the component K_{11} and some similarities for both matrices can be noticed, however for K_{33} the MLPT Green's matrix has a fewer number of real singularities. Note that with increasing frequency-thickness the difference between the values of Green's matrix components for MLPT and elasticity theory grows.

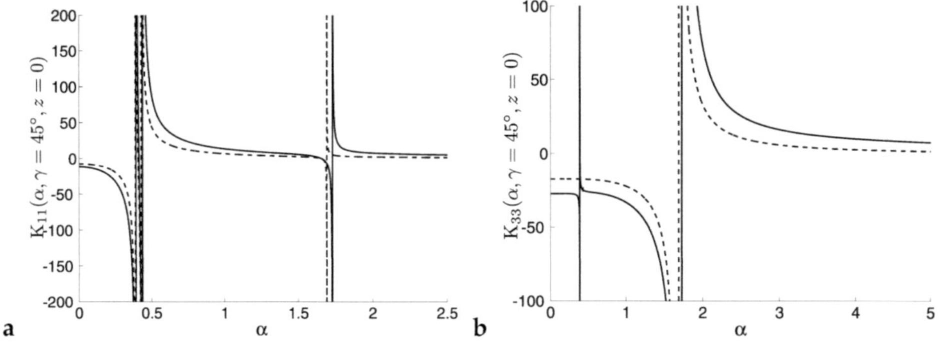

Figure 3.4: Comparison between Green's matrix components K_{11} (a) and K_{33} (b) for the elastodynamic problem ("$-$") and MLPT ("$--$") in dependence on α for $[0/90]_s$ laminated composite plate at frequency-thickness $f \cdot h = 300\ \mathrm{KHz} \cdot \mathrm{mm}$ for $\gamma = 45°$

Application of the integral approach to the equation of the CLPT (2.36) yields

$$\mathbf{T}_C \mathbf{U}_C = V(\omega)\mathbf{Q} \tag{3.76}$$

with $\mathbf{U}_C(\alpha_1, \alpha_2, \omega) = (U_0, V_0, W_0)^T$ and with non-symmetric but positive defninite Hermitian matrix $\mathbf{T}_C$. Green's matrix of the Classical Laminated Plate is then computed as follows:

$$
\begin{aligned}
\mathbf{U}_{\text{CLPT}}(\alpha_1, \alpha_2, z, \omega) &=
\begin{pmatrix}
U_{0,1} - z W_{0,1} & U_{0,2} - z W_{0,2} & U_{0,3} - z W_{0,3,} \\
V_{0,1} - z W_{0,1} & V_{0,2} - z W_{0,2} & V_{0,3} - z W_{0,3,} \\
W_{0,1} & W_{0,2} & W_{0,3}
\end{pmatrix}
\mathbf{Q}(\alpha_1, \alpha_2) \\
&= \mathbf{K}_{\text{CLPT}}(\alpha_1, \alpha_2, z, \omega)\mathbf{Q}(\alpha_1, \alpha_2).
\end{aligned}
\tag{3.77}
$$

The expressions obtained for models based on plate theory are much simpler and allow a much quicker evaluation of Green's matrix of the wave propagation problem in comparison to the algorithm described in section 3.2.1. However, as mentioned in section 2.2.3 and as will be shown by numerical examples, the application of plate theories is limited in practice to the case of low excitation frequencies.

The properties of Green's matrices for MLPT and CLPT are generally similar to the properties of Green's matrix of the elastodynamic problem (section 3.2.1). For example, the values of Green's matrix components are real or pure imaginary for $\mathrm{Re}\,\alpha = \alpha$ and the matrices $\mathbf{K}_{\text{MLPT}}$ and $\mathbf{K}_{\text{CLPT}}$ have the same symmetric properties (3.68) as the matrix $\mathbf{K}$ in (3.63). Moreover, if the symmetric plate is considered, both models give the equation with the in-plane and out-of-plane motions independently of each other, i.e. the motions are decoupled.

However, some differences are presented. The asymptotic properties of Green's matrices for MLPT and CLPT are obtained for $\alpha \to \infty$ as follows:

$$
\mathbf{K}_{\text{CLPT}} \sim \frac{1}{\alpha^2}
\begin{pmatrix}
1 & 1 & \alpha^{-1} \\
1 & 1 & \alpha^{-1} \\
\alpha^{-1} & \alpha^{-1} & \alpha^{-2}
\end{pmatrix},
\quad
\mathbf{K}_{\text{MLPT}} \sim \frac{1}{\alpha^2}
\begin{pmatrix}
1 & 1 & \alpha^{-1} \\
1 & 1 & \alpha^{-1} \\
\alpha^{-1} & \alpha^{-1} & 1
\end{pmatrix}.
\tag{3.78}
$$

The influence of the properties of Green's matrices described here on the algorith of computation of the corresponding wave fields is discussed in the following chapters.

4 Dispersion properties of laminated composite plates

4.1 Dispersion equation

This part represents the algorithm of computation of the poles of Green's matrix in the wavenumber-frequency domain. This algorithm is used for studying the dispersion properties of various laminated composites.

4.1.1 Dispersion equation for elastodynamic problem

As it is stated in section 3.2.3 due to the property (3.66) all poles of Green's matrix are defined by the characteristic (or dispersion) equation

$$\det \mathbf{W}(\alpha, \gamma, \omega) = 0. \tag{4.1}$$

This equation has to be solved for the values α in dependence on γ and ω as $k(\gamma, \omega)$. For each pair of the parameters γ and ω, there exist a finite number of real roots and a countable number of complex roots. Note that if $k(\gamma, \omega)$ satisfies Equation (4.1), its value taken with the opposite sign $-k(\gamma, \omega)$ will also satisfy (4.1). This results in a periodicity of the poles with respect to π due to (3.69). In addition, for a complex root k of the dispersion equation, the values $-k$, k^* and $-k^*$, where "$*$" denotes the complex conjugate, are also satisfying the dispersion equation, i.e. while root-finding it is enough to find the roots with non-negative both real and imaginary parts, i.e. k: $\operatorname{Re} k \geq 0$, $\operatorname{Im} k \geq 0$. The roots of the characteristic equation (4.1) can be represented in dependence on parameters γ and ω by the continuous surfaces (or curves), called dispersion curves.

The problem of finding the dispersion surfaces was investigated by many authors using transfer and global matrix techniques (see section 2.3.2). The dispersion curves found from (4.1) should coincide with the curves found using transfer and global matrix techniques. Before starting with a description of the root-finding (or pole-finding) algorithm, some properties of (4.1) need to be discussed.

In particularly, some roots of the dispersion equation found for fixed γ and ω are eliminable poles of Green's matrix, i.e. at the same time they are zeros of the same degree of all components $\tilde{K}_{ij}(\alpha_1, \alpha_2, z, \omega)$ in (3.66) and do not define any discontinuities of Green's matrix components and should be excluded from the analysis. All other roots of (4.1) are the poles of Green's matrix, i.e. if $k(\gamma, \omega)$ is the pole of the order m of Green's matrix[1], it follows that

$$\lim_{\alpha \to k(\gamma,\omega)} K_{ij}(\alpha, \gamma, z, \omega)\,(\alpha - k(\gamma, \omega))^m \neq 0 \tag{4.2}$$

for at least one component of Green's matrix. Equation (4.2) can be used to determine if the root of the dispersion equation found is eliminable, however it is more convenient to use Cauchy's argument principle [3]

$$\int_C \frac{f'(\alpha)}{f(\alpha)}\, \mathrm{d}\alpha = 2\pi \mathrm{i}(N - P), \tag{4.3}$$

where C is some closed contour, N and P denote respectively the number of zeros and poles of meromorphic function $f(\alpha)$ inside the contour C. The zeros (poles) are counted as many times as their multiplicity (order). As an integration contour C a circle of small radius with the center in a point of the complex plane can be chosen, which is necessary for checking up, whether it is a pole of Green's matrix or not.

Another problem consists in the numerical calculation of eigenvectors $\mathbf{m}_j$ in (3.52) for each ply in a plate. Due to the fact that the vector $-\mathbf{m}_j$ is also an eigenvector of the problem and replacing the vector $\mathbf{m}_j$ by $-\mathbf{m}_j$ in (3.52), the sign of determinant will be changed to the opposite one, i.e. the left side of Equation (4.1) becomes to be $-\det \mathbf{W}(\alpha, \gamma, \omega)$. This concludes that only the absolute value of function $\det \mathbf{W}(\alpha, \gamma, \omega)$ can be computed numerically stable as a continuous function of the parameters γ and ω. It follows that the usual bisection method [134], Newton's [134] and Mullers [93] methods are not suitable for root-finding. Instead of them, the non-gradient optimization algorithms not requiring a continuous first derivative of the goal function can be applied. However, they are usually slower than aforementioned root-finding algorithms. Due to this fact and due to the presence of eliminable roots it is more convenient to consider the equation

$$\frac{1}{\det \mathbf{K}(\alpha, \gamma, z = 0, \omega)} = 0 \tag{4.4}$$

instead of the dispersion equation (4.1), where $\mathbf{K}(\alpha, \gamma, z = 0, \omega)$ is Green's matrix computed at the surface of the structure $z = 0$. Evidently, the poles of Green's matrix satisfy (4.4). This equation can be replaced by another equation

$$\frac{1}{K_{11}(\alpha, \gamma, z = 0, \omega) + K_{22}(\alpha, \gamma, z = 0, \omega) + K_{33}(\alpha, \gamma, z = 0, \omega)} = 0 \tag{4.5}$$

---

[1]Note that if a layered half-space is considered instead of the plate, also the discontinuities of order $1/2$ are presented, i.e. *branch points* [9, 10].

with the same roots. Both dispersion equations (4.4) and (4.5) follow from the property of Green's matrix components given by (3.66).

Note that some of the discontinuities are presented not for all components of Green's matrix and it is not sufficient to consider only one component of Green's matrix $K_{ij}^{-1}(\alpha, \gamma, \omega) = 0$ for the determination of all poles. The main disavantage of both Equations (4.4), (4.5) consists in a presence of discontinuties in the functions, the zeros of which should be found. It brings additional difficulties into the numerical implementation of the pole-finding algorithm.

Another difficulty observed is the intersection of the dispersion curves corresponding to different wave modes, which can distort the following of the curve while changing angle γ or frequency ω. However, in case of a laminated plate symmetric with respect to mid-plane with all plies of the elastic symmetry not lower than orthotropic, the symmetric and antisymmetric wave modes can be decoupled and the root-finding is simplified. It can be done by considering only its upper half consisting of $N/2$ layers instead of the whole laminate and applying to the lower boundary of laminate's half the boundary conditions given by (2.48), (2.49). The matrix $\mathbf{S}^{N/2}$ for lower boundary condition in the wavenumber-frequency domain is then obtained for symmetric and antisymmetric wave modes as follows[1]:

$$\mathbf{S}_S^{N/2} = \begin{pmatrix} 0 & 0 & 0 & C_{55} & C_{54} & 0 \\ 0 & 0 & 0 & C_{45} & C_{44} & 0 \\ 0 & 0 & 1 & 0 & 0 & 0 \end{pmatrix},$$

$$\mathbf{S}_A^{N/2} = \begin{pmatrix} 1 & 0 & 0 & 0 & 0 & 0 \\ 0 & 1 & 0 & 0 & 0 & 0 \\ 0 & 0 & -i(\alpha_2 C_{34} + \alpha_1 C_{35}) & 0 & 0 & C_{33} \end{pmatrix}. \tag{4.6}$$

Considering the half of the laminate with lower boundary conditions given by matrices $\mathbf{S}_S^{N/2}$ or $\mathbf{S}_A^{N/2}$, the characteristic equations of symmetric and antisymmetric wave modes are obtained as

$$\det \mathbf{W}_S(\alpha, \gamma, \omega) = 0 \quad \text{or} \quad \det \mathbf{W}_A(\alpha, \gamma, \omega) = 0. \tag{4.7}$$

Similar equations can be obtained if the transfer and global matrix algorithms are applied (see section 2.3.2). The use of Equations (4.7) allows to reduce the computational time due to the lower number of layers in each composite and to study the dispersion properties of symmetric and antisymmetric wave modes separately avoiding the situation of the curves intersection, while the curves of one type (either symmetric or antisymmetric) do not cross each other.

---

[1]Some components are zero due to an assumption of at least orthotropic properties of layers and due to the boundary conditions (2.48), (2.49).

4.1.2 Dispersion equation for CLPT and MLPT

In case of modelling of a structure using plate theories, the dispersion equation is given by setting equal to zero the determinants of the matrices $\mathbf{T}_C$ and $\mathbf{T}_M$ for CLPT and MLPT, respectively,

$$\det \mathbf{T}_M(\alpha, \gamma, \omega) = 0, \tag{4.8}$$

$$\det \mathbf{T}_C(\alpha, \gamma, \omega) = 0, \tag{4.9}$$

as it follows from (3.73) and (3.76). As the dimension of the matrices $\mathbf{T}_C$ and $\mathbf{T}_M$ is lower than for elastodynamic problem and due to the calculation of their components in the frequency-wavenumber domain as analytical functions, the roots of the dispersion equations (4.8), (4.9) are computed essentially faster than for the model based on elastodynamic equations.

As well as in case of modelling by means of the elasticity theory, the problem of root-finding for the characteristic equation becomes simpler in case of symmetry of a composite plate with respect to its mid-plane. For CLPT due to zero coupling components $\mathbf{B} = 0$ and $I_1 = 0$ in matrix $\mathbf{T}_C$, the equations for the determination of wavenumbers of symmetric wave modes become to be

$$T_{C,11}(\alpha, \gamma, \omega) T_{C,22}(\alpha, \gamma, \omega) - T_{C,12}^2(\alpha, \gamma, \omega) = 0. \tag{4.10}$$

The corresponding equation for determination of wavenumbers of antisymmetric wave mode is given by

$$T_{C,33}(\alpha, \gamma, \omega) = 0, \tag{4.11}$$

or in an explicit form by (2.62). The Equations (4.10), (4.11) unlike (4.7) given previously have only a final number of roots at any frequency ω and angle γ. Moreover, the expression on the left-side of (4.10) is a biquadratic polynomial with respect to α. It has only two pairs of real roots $\pm k_1(\gamma, \omega)$ and $\pm k_2(\gamma, \omega)$. The roots $k_1(\gamma, \omega)$ and $k_2(\gamma, \omega)$ correspond to the fundamental symmetric modes S_0 and SH_0. Furthermore, the wavenumbers of these modes are depending linearly on the frequency ω and the corresponding phase and group velocities are independent of the frequency. The wave modes resulting from the equation of antisymmetric motion (4.11) are discussed previously in section 2.3.6.3.

Considering the symmetric motion using MLPT, it is obtained that the characteristic equation is the same as for CLPT due to $T_{M,ij} = T_{C,ij}$ for $i, j = 1, 2$:

$$\begin{aligned}
T_{M,11}(\alpha, \gamma, \omega) T_{M,22}(\alpha, \gamma, \omega) - T_{M,12}^2(\alpha, \gamma, \omega) = \\
T_{C,11}(\alpha, \gamma, \omega) T_{C,22}(\alpha, \gamma, \omega) - T_{C,12}^2(\alpha, \gamma, \omega) = 0.
\end{aligned} \tag{4.12}$$

The dispersion equation for the antisymmetric motion obtained using MLPT differs from (4.11) obtained previously as follows:

$$\begin{vmatrix} T_{M,33}(\alpha,\gamma,\omega) & T_{M,34}(\alpha,\gamma,\omega) & T_{M,35}(\alpha,\gamma,\omega) \\ -T_{M,34}(\alpha,\gamma,\omega) & T_{M,44}(\alpha,\gamma,\omega) & T_{M,45}(\alpha,\gamma,\omega) \\ -T_{M,35}(\alpha,\gamma,\omega) & T_{M,45}(\alpha,\gamma,\omega) & T_{M,55}(\alpha,\gamma,\omega) \end{vmatrix} = 0. \tag{4.13}$$

The determinant on the left side can be represented as a polynomial of degree 6 with respect to α, which has three pairs of roots corresponding to antisymmetric modes A_0, A_1 and SH_1. The wavenumbers of the mode A_0 are always real, in contrast to the wavenumbers of modes A_1 and SH_1, which are pure imaginary at low frequencies and become to be real for frequencies higher than the cut-off frequency. The movement of the values of wavenumbers of A_1 and SH_1 in a complex plane with increasing the frequency is shown in Figure 4.1a.

The numerical computation of the roots of dispersion equations for a modelling based on laminated plate theories in comparison with models using elastodynamic equations is much faster. The functions in the equations are given analytically, do not have any singularities and all roots of the characteristic equations are not eliminable discontinuities of Green's matrix components[1]. Furthermore, for the symmetric composite plate standard root-finding algorithms like bisection method or Muller's method can be used.

4.2 Algorithm for calculation of wavenumbers

The properties of roots of the characteristic equations described in section 4.1.1 require a development of a special stable algorithm based on standard root-finding algorithms and taking into account properties of the equations under study. Such an algorithm developed by the author of this thesis and implemented for the investigation of dispersion of composites in the works [60, 61] is presented in this section. This algorithm is based on Muller's method [93] and employs the computation of wavenumbers for some initial values of parameters γ and ω and then realizes the curve following which starts from the previously found solution. Muller's method is chosen because it does not need the direct calculation of derivative and because of its high speed of convergence for functions of complex variables. Note that the algorithms for obtaining the wavenumbers of isotropic [82] and anisotropic [43, 69] layered structures are similar to the algorithm presented here, however the algorithm presented here involves some important improvements.

---

[1]Note that due to decoupling of the problem the symmetric wave modes do not influence the vertical displacement.

Below the steps of the algorithm of calculating the poles of Green's matrix for angles $\gamma \in [0, 2\pi]$ and frequencies $\omega \in [\omega_{min}, \omega_{max}]$ are presented[1]. The poles are searched with accuracy ε, the parameters γ and ω are incremented by the values $\Delta\gamma$ and $\Delta\omega$ correspondingly.

1. First, the algorithm checks if the laminated plate is symmetric. If it is true, the wave motion can be decoupled, the following steps will be done for antisymmetric and symmetric modes separately. Otherwise, one of the characteristic equations (4.4), (4.5) or (4.7) is used. The values of starting frequency ω_{max} and the angle $\gamma = 0$ are chosen. The counters n_γ and n_ω, which describe the number of completed steps of the algorithm with respect to variables γ and ω respectively, are both set to zero.

2. If $n_\omega = 0$, all poles with respect to α in domain $Y : 0 \le \mathrm{Re}\,\alpha \le \alpha_{Re}$, $0 \le \mathrm{Im}\,\alpha \le \alpha_{Im}$ are searched for the frequency ω_{max} and $\gamma = 0$ using Muller's method with accuracy given by ε and initial approximations given as points in the domain Y as $\alpha_{jm} = (j-1)\alpha_{Re}/N_{Re} + \mathrm{i}(m-1)\alpha_{Im}/N_{Im}$, where $j = 1, \ldots, N_{Re} + 1$ and $m = 1, \ldots, N_{Im} + 1$. Note that the values α_{Re} and α_{Im} describe the pre-defined limitations on the searching area, N_{Re} and N_{Im} correspond to the given number of discrete points taken along the real and imaginary axis to get the initial approximations in the domain of search with a total number of $(N_{Re} + 1) \cdot (N_{Im} + 1)$ initial approximations. Optimal values are estimated on various numerical examples at frequency-thickness products[2] $f \cdot h < 1.5\ \mathrm{MHz} \cdot \mathrm{mm}$ to be $N_{Re} = 50$, $N_{Im} = 20$, $\alpha_{Re} = 10$ and $\alpha_{Im} = 3$. If $n_\omega > 0$, the values of poles found for $\gamma = 0$ and frequency $\omega + \Delta\omega\,(n_\omega - 1)$ are used as initial approximations instead of α_{jm}. Only unique[3] values of roots found are considered as roots of equation.

Hereinafter the values are checked to be the poles of Green's matrix by numerical evaluation of the contour integral (4.3), i.e. for the value $k = k(\gamma, \omega)$ the contour integral over the circle C_k with the center at k of radius 5ε is computed using the following formula based on the midpoint approximation of definite integral and central difference for derivative

$$P \approx -\frac{1}{2\pi \mathrm{i}} \sum_{j=1}^{N-1} \frac{f(k + 5\varepsilon\,\mathrm{e}^{\mathrm{i}\beta}, \gamma, \omega)\big|_{\beta=2\pi j/N} - f(k + 5\varepsilon\,\mathrm{e}^{\mathrm{i}\beta}, \gamma, \omega)\big|_{\beta=2\pi(j-1)/N}}{f(k + 5\varepsilon\,\mathrm{e}^{\mathrm{i}\beta}, \gamma, \omega)\big|_{\beta=2\pi(j-1/2)/N}}, \quad (4.14)$$

where the determinant of Green's matrix $f = \det \mathbf{K}(\alpha, \gamma, z, \omega)$ or its components $K_{ij}(\alpha, \gamma, z, \omega)$ can be used as the function $f(\alpha, \gamma, \omega)$. The value P corresponds to the order of the pole k. Note that the formula assumes that no zeros or other

---

[1] This algorithm requires that all values are given in dimensionless form, i.e. instead of ω the value $\bar{\omega}$ is used.

[2] At higher frequencies the limits along the real-axis should be higher since the real part of wavenumbers increases as frequency increases.

[3] The roots of opposite sign are considered as a single root.

poles are located inside of the region bounded by circle C_k. Usually the number of $N = 8 \ldots 32$ is sufficient to get the true value of P after rounding.

3. The angle γ is changed to $\gamma + \Delta\gamma$, the value of n_γ is increased by 1. If the value of $\gamma \leq 2\pi$, the next step is 4, else the algorithm proceeds with step 5.

4. For the new value of γ the initial approximations of the poles can be found by extrapolating the values of wavenumbers found at previous steps with respect to γ on current step. Depending on the number of the steps with respect to γ n_γ previously completed for the current frequency ω, the initial approximations for Muller's method can be extrapolated from previous steps as[1]

$$
\begin{aligned}
k(\gamma + \Delta\gamma) &= k(\gamma), \quad \text{for } n_\gamma = 1, \\
k(\gamma + \Delta\gamma) &= 2k(\gamma) - k(\gamma - \Delta\gamma), \quad \text{for } 1 < n_\gamma < 6, \\
k(\gamma + \Delta\gamma) &= 3k(\gamma - \Delta\gamma) - 3k(\gamma - 3\Delta\gamma) + k(\gamma - 5\Delta\gamma), \quad \text{for } n_\gamma \geq 6.
\end{aligned}
\tag{4.15}
$$

The use of the extrapolation formulas (4.15) not only gives a better convergence to the roots of the dispersion equation, but also minimizes the risk of following the wrong curve if two dispersion curves cross, because in a quadratic extrapolation applied if $n_\gamma \geq 6$ the alternate points instead of the consecutive points are used, that delays the influence of any erroneous points by one step [82].

Then all roots of dispersion equation are searched. To prevent the situation that two different initial approximations give the same root, the root k already found can be eliminated by considering the function $f(\alpha)(\alpha - k)$ instead of function $f(\alpha)$. Hence, sometimes the algorithm springs to another dispersion curve k_j which was not followed before. If this situation occurs, the algorithm tries to find the missed root considering another dispersion equation as chosen initially. Usually this allows to find the root correctly. If it does not help, the root k_j is eliminated[2] and again the previous initial approximation is used. After finding all poles for the angle γ, step 3 is evaluated.

5. If $\omega - \Delta\omega > \omega_{\min}$, the new frequency is set to be $\omega = \omega - \Delta\omega$, the value n_ω is increased by 1, n_γ is set to 0 and step 2 is evaluated, else the algorithm terminates.

Note that step 2 takes long computational time, because the good initial approximations for roots are unknown and many initial approximations are used to ensure that no poles located in the domain Y are missed. All other steps are taking less computational time due to the use of good initial approximations. The increments $\Delta\gamma$ and $\Delta\omega$ recommended for numerical computations are $\Delta\gamma \sim 0.0005\pi \ldots \pi/180$ and $\Delta\omega \sim 0.00001 \ldots 0.02$. The computational time of the algorithm depends mostly on the number of layers in a composite plate, the number of roots followed by the algorithm and the values of increments $\Delta\gamma$ and $\Delta\omega$.

---

[1]Dependence of the roots on ω is omitted here for simplicity.
[2]This root is not followed after this iteration.

Remark 4.1 *For low frequencies the wavenumbers of three fundamental wave modes are all approaching zero, the differences between their values are small and sometimes the algorithm misses one of the roots. However, the values of phase velocities $c_p = \omega/k$ of the fundamental modes are different, and instead of wavenumbers the values of c_p for each mode are searched.*

Remark 4.2 *The algorithm can be modified by using the extrapolation formulas similar to (4.15) for calculating the initial approximations on the next iteration with respect to ω.*

Remark 4.3 *The same algorithm with some minor modifications can also be used for finding the roots of the dispersion equation of plate theories (4.8), (4.9).*

4.3 Investigation of dispersion properties of laminated composites

The methods of calculation of dispersion curves for laminated composites described previously in this chapter are applied in this section for studying the dispersion properties of various composite plates. The dispersion curves of wave modes with real-valued wavenumbers are of most interest since they correspond to propagating (non-attenuating) Lamb waves. Below the properties of Lamb waves observed in the composite specimens under study are discussed.

4.3.1 Common behaviour of dispersion curves increasing the frequency

As mentioned in section 2.3.4, in a plate with a stress-free lower boundary at least three non-attenuating waves are observed, so called fundamental wave modes A_0, S_0 and SH_0. These modes correspond to Green's matrix poles of first order (except of the frequency $\omega = 0$, i.e. for the static case). Unlike the fundamental modes, most of the other so called *higher-order* wave modes become to have real-valued wavenumbers only for frequencies higher than their cut-off frequencies. At cut-off frequency the corresponding wavenumber curve passes through the origin of the complex plane α (point 3 in Figure 4.1a). The cut-off frequency is reached by the wave mode simultaneously in all directions γ. For frequencies lower than a cut-off frequency (points 1 and 2 in Figure 4.1a) the wavenumbers of the corresponding mode are pure imaginary, for frequencies higher (points 4 and 5 in Figure 4.1a) than a cut-off frequency the wavenumbers are real-valued. At the cut-off frequency, the wave mode corresponds to Green's matrix pole of second order, at all other frequencies the corresponding poles are of first order. However, some laminated plates similarly to isotropic plates have wave modes with complex-valued wavenumbers, for which the imaginary parts become to be small while increasing the frequency (points 1 and 2 in Figure 4.1b) and at some frequency, the wavenumbers become to be real (points 4 and 5 in Figure 4.1b). In the sequel the

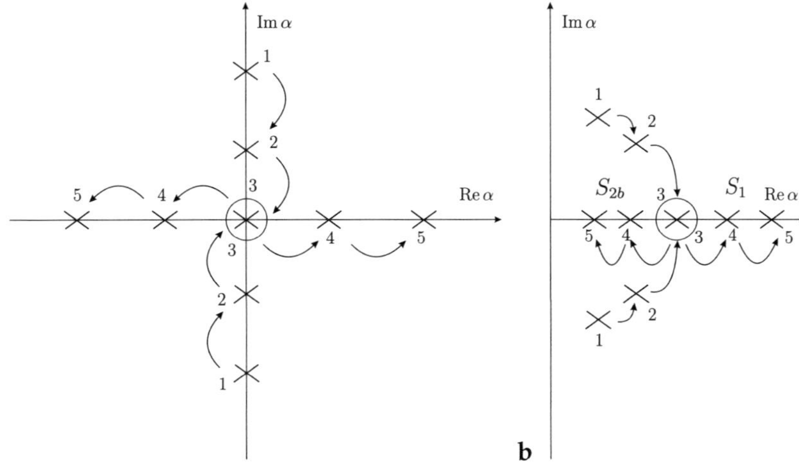

Figure 4.1: (a) Pure imaginary poles approaching the real axis while the frequency increases. (b) Complex poles approaching the real axis while the frequency increases and producing the backward propagating mode S_{2b}

frequency at which the propagating wave mode firstly observed is called a *frequency of first appearance* of the wave mode. Since the complex conjugate values of wavenumbers also satisfy the dispersion equation (see points denoted as 1 and 2 in Figure 4.1b), the order of Green's matrix pole corresponding to this frequency is 2. Note that in different directions γ, this frequency is reached by the wave modes not simultaneously, i.e. the mode is still non-propagating for some angles of γ while for other angles it has already real-valued wavenumbers. For frequencies higher than the frequency of first appearance of the mode, there are two propagating wave modes observed. One of the two modes, e.g. S_1 in Figure 4.1b, has normal dispersion (positive group velocity) and behaves like other higher-order wave modes at higher frequencies. The wavenumbers of the second mode decrease with increasing of frequency, e.g. S_{2b} in Figure 4.1b. The group velocity corresponding to this mode is negative and the corresponding poles of Green's matrix are *irregular*. At some frequency, its wavenumber curve passes through the origin and takes pure imaginary values. The corresponding frequency is a cut-off frequency and is reached by the wave mode for all angles γ simultaneously. Further following of this curve shows that for higher frequencies this mode behaves as all other higher-order wave modes with pure imaginary wavenumbers, and can become to be propagating at some cut-off frequency.

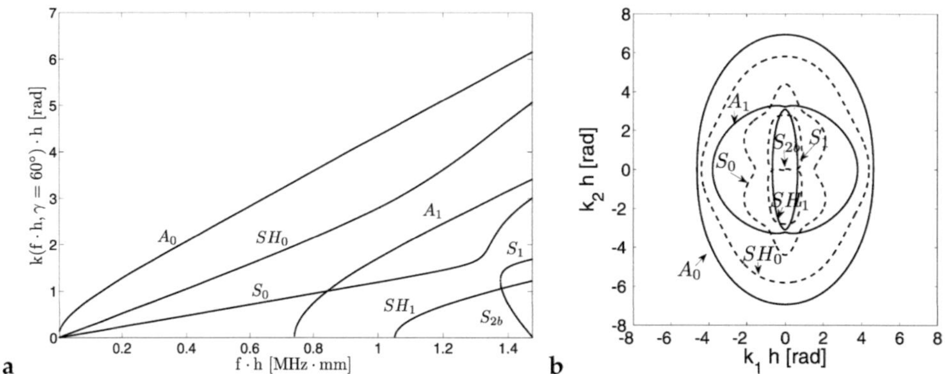

Figure 4.2: Wavenumber curves of the poles in unidirectional plate of graphite-epoxy I when $\gamma = 60°$ depending on frequency-thickness product $f \cdot h$ (a) and depending on γ when $f \cdot h = 1.472$ MHz $\cdot$ mm (b)

An example of the wavenumber curves calculated in dependence on frequency for fixed angle $\gamma = 60°$ is presented in Figure 4.2a for a unidirectional plate made of graphite-epoxy I (Table A.1 in Appendix A.9). For frequency-thicknesses lower than the first cut-off frequency-thickness $f \cdot h \approx 750$ KHz $\cdot$ mm, only three waves are propagating. For $750 < f \cdot h < 1380$ (in KHz $\cdot$ mm) in addition to the fundamental wave modes two higher-order wave modes - A_1 and SH_1 - are observed. For the frequency-thickness $f \cdot h \approx 1380$ KHz $\cdot$ mm two additional waves become to have real wavenumbers. The wavenumbers of one of these two higher-order wave modes grow with increasing frequency, in constrast, the wavenumbers of the second of these two wave modes decline with increasing frequency, i.e. $dk(\omega)/d\omega = 1/c_g < 0$ and this mode is a backward mode corresponding to an irregular pole of Green's matrix. At $f \cdot h \approx 1500$ KHz $\cdot$ mm, the curve of its mode crosses the origin and for frequencies higher than this cut-off frequency, this mode has pure imaginary wavenumbers[1]. Analysing the wavenumber curves, the intersections between the curves are observed, e.g. at $f \cdot h \approx 830$ KHz $\cdot$ mm and $f \cdot h \approx 1390$ KHz $\cdot$ mm. However, the intersecting curves correspond to two different types of wave modes - antisymmetric and symmetric ones. In Figure 4.2b the wavenumbers of antisymmetric (red lines) and symmetric (dashed blue lines) are plotted in dependence on angle γ if the frequency-thickness product is given by $f \cdot h = 1473$ KHz $\cdot$ mm. This figure illustrates well the complicated dependence of the wavenumbers on direction γ due to the high influence of anisotropy of the plate.

The behaviours displayed in Figure 4.1 are also valid for multilayered composite plates. The corresponding numerical results are illustrated in the next subsection.

---

[1]However, usually this mode has a second cut-off frequency, at which it becomes to be propagating with wavenumbers corresponding to regular poles of Green's matrix.

4.3.2 Numerical results

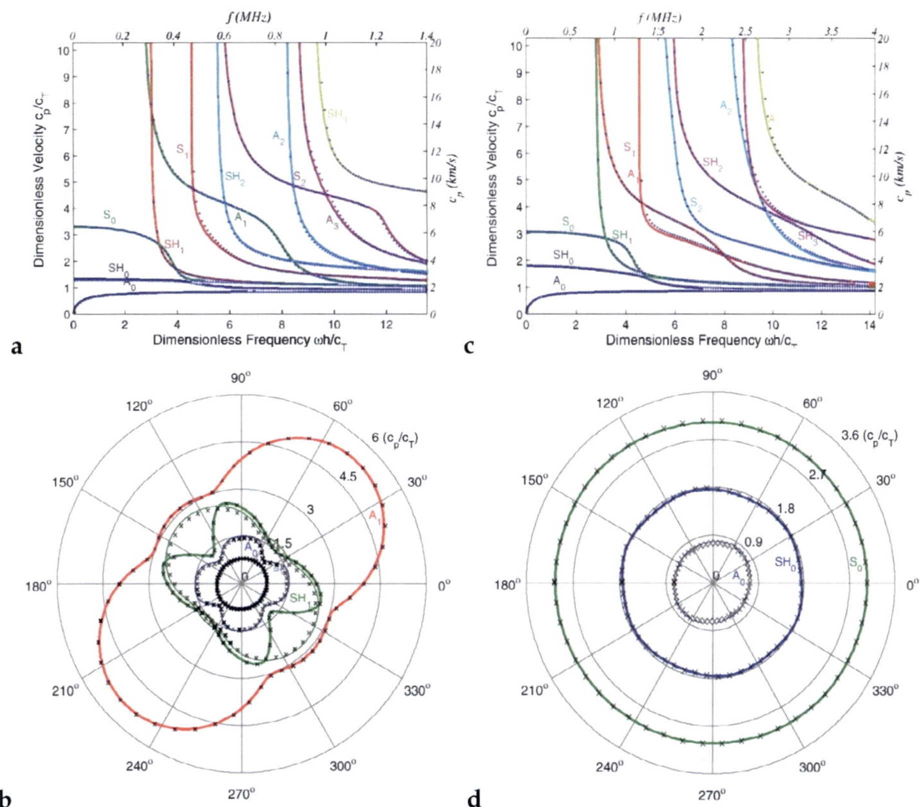

Figure 4.3: Graphs of phase velocities (marked with asterisks) of Lamb waves in comparison to results (grid lines) published in [149] for a $[45_6/-45_6]_s$ laminate in direction of $\gamma = 30°$ (a) and for a $[45/-45/0/90]_s$ laminate in direction of $\gamma = 45°$ (c), for a $[45_6/-45_6]_s$ composite plate (b) at dimensionless frequency $\omega h/c_T = 4$ ($f \cdot h = 1238$ KHz $\cdot$ mm) and for a $[45/-45/0/90]_s$ composite plate (d) at dimensionless frequency $\omega h/c_T = 1.78$ ($f \cdot h = 551$ KHz $\cdot$ mm)

In order to validate the method presented previously, the dispersion curves obtained by its application are compared with the dispersion curves obtained by the transfer matrix method used by the authors in [149]. The results are compared for the cross-ply laminated composite plate consisting of 24 layers with stacking sequence[1] $[45_6/-45_6]_s$

---

[1]Due to the continuity of the stresses and displacements at layer interfaces, this stacking sequence can

and for the quasi-isotropic laminated composite plate consisting of 8 layers with stacking sequence $[45/-45/0/90]_s$. All layers in both laminated plates are made of the orthotropic material AS4/3502, the properties of which are given in Table A.1 in Appendix A.9.

Figures 4.3a, b, c and d show the dispersion curves[1] (marked with asterisks) obtained by the authors and superimposed on the corresponding graphs from article [149]. These curves are the dimensionless velocities c_p/c_T of propagation of symmetric and antisymmetric modes for the dimensionless frequency $\omega = \omega h/c_T$. The value of the reference velocity c_T for normalization of values is defined as $c_T = \sqrt{G_{12}/\varrho} = 1945$ m/s, where $G_{12} = 5.97 \cdot 10^9$ Pa is the shear modulus in the xy plane and ϱ is the density of AS4/3502 material. Figure 4.3a corresponds to the phase velocities c_p (defined in (2.54)) in the wavenumber domain of fundamental and higher-order wave modes in $[45_6/-45_6]_s$ composite at an angle $\gamma = 30°$. In Figure 4.3c the phase velocities c_p of Lamb wave modes are plotted for the $[45/-45/0/90]_s$ composite plate and an angle $\gamma = 45°$. Figures 4.3b and d present the graphs of phase velocities depending on the incident angle γ for the dimensionless frequency $\omega h/c_T = 4$ (frequency-thickness product $f \cdot h = 1238$ KHz $\cdot$ mm) (b) for $[45_6/-45_6]_s$ and $\omega h/c_T = 1.78$ (frequency-thickness product $f \cdot h = 551$ KHz $\cdot$ mm) for $[45/-45/0/90]_s$ composite plates, respectively. As may be seen from the given figures, the curves obtained using the methods given in this work showed the qualitative agreement of the results with the results published in [149]; the inaccuracies are caused by the imperfect superimposition of plots. Thus, the algorithm of calculation of the wavenumbers, presented in this thesis, is valid for the study of dispersion properties of laminated composite plates.

Next, the results obtained using a model based on the elasticity theory and the laminated plate theories are compared. As already mentioned previously, the plate theories give results coinciding with the results obtained using an elasticity theory only in range of low frequencies. It is well demonstrated in Figure 4.4, which represents the wavenumber curves obtained for $f \cdot h \leq 200$ KHz $\cdot$ mm applying the algorithm of calculation of the roots of the dispersion equations for elasticity theory (ET) and both plate theories - CLPT and MLPT. The results are compared in directions $\gamma = 0°$ (Figure 4.4a and c) and $\gamma = 45°$ (Figure 4.4b and d) for a laminated plate of graphite-epoxy II (Table A.1 in Appendix A.9) with stacking sequence $[0/90/45/-45]_s$. There are no differences observed between the dispersion curves of S_0 and SH_0 obtained for different models at low frequencies, however the difference in the modelling of laminated

---

be simplified to $[45/-45]_s$.

[1]Note that the colours in these figures are confusing due to the use of one colour for representing curves of different wave modes, e.g. red-coloured curve in figure 4.3a corresponds to SH_1 whereas in figure 4.3b it corresponds to A_1. However, these colours are taken from [149].

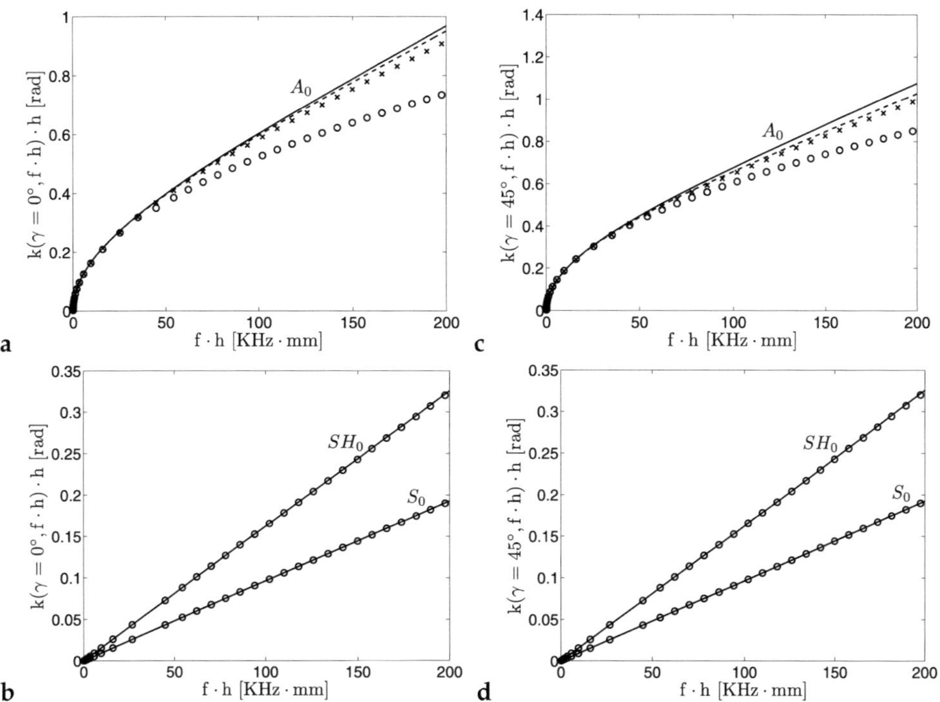

Figure 4.4: Wavenumbers k of fundamental Lamb wave modes A_0 (a and b) and S_0, SH_0 (c and d) at frequency-thickness products $f \cdot h \leq 200$ KHz $\cdot$ mm in a laminated plate of graphite-epoxy II with stacking sequence $[0/90/45/ - 45]_s$ in fixed directions of $\gamma = 0°$ (a,c) and $\gamma = 45°$ (b,d)

plate becomes to be clearly visible if the curves of A_0 are analysed. Furthermore, this difference grows with increasing frequency.

The differences between the modelling approaches are well displayed by the curves of the group velocity in dependence on observation direction φ (see Equation (2.55)[1]), i.e. the group velocity of wave front (GWS), for the A_0 wave mode. These velocities are calculated for the A_0 mode using the roots of dispersion equations (4.9) for CLPT, (4.8) for MLPT and (4.1) for elasticity theory and plotted for two values of frequency-thickness products $f \cdot h = 11$ KHz $\cdot$ mm and $f \cdot h = 50$ KHz $\cdot$ mm in dependence on φ in Figures 4.5a and b, respectively. While all these curves are well coinciding for the frequency-thickness product value of 11 KHz $\cdot$ mm, the differences between dispersion curves obtained using MLPT and elasticity theory at the higher frequency-thickness

---

[1]This formula provides a way for the computation of these group velocities in an implicit dependence on φ.

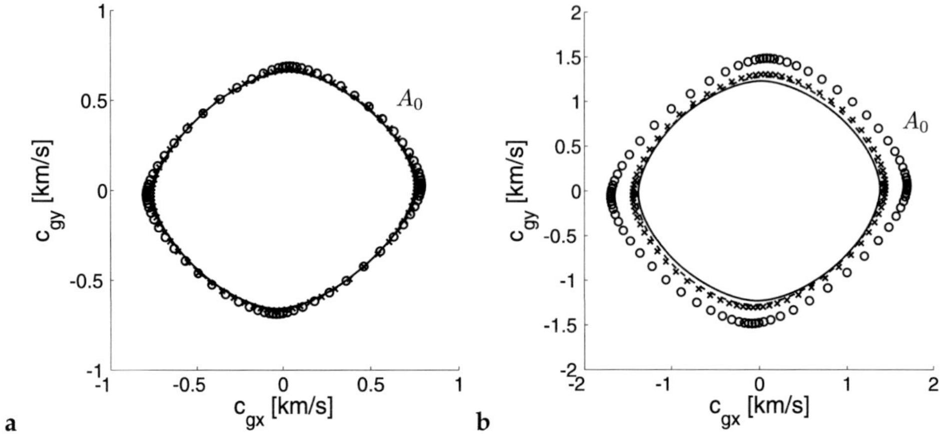

Figure 4.5: Group wave surfaces (GWS) of fundamental Lamb wave mode A_0 at frequency-thickness products $f \cdot h = 11$ KHz $\cdot$ mm (a) and $f \cdot h = 50$ KHz $\cdot$ mm (b) in a laminated plate of graphite-epoxy II with stacking sequence $[0/90/45/-45]_s$

product value of 50 KHz $\cdot$ mm are notable. The group velocities of the wave front for the mode A_0 obtained using CLPT (in Figure 4.5 marked as "o") are considerably higher than the values obtained using MLPT and elasticity theory for the frequency-thickness product value of 50 KHz $\cdot$ mm.

The comparison of the group velocities of wave fronts of fundamental wave modes for the composite plate of IM7-Cycom-977-3 (Table A.1 in Appendix A.9) with stacking sequence $[0/45/-45/90]_{2s}$ is presented for two frequency-thickness products $f \cdot h = 11$ KHz $\cdot$ mm (Figures 4.6a and b) and $f \cdot h = 500$ KHz $\cdot$ mm (Figures 4.6c and d). The coincidence of the dispersion properties of symmetric wave modes observed at low frequencies previously is observed also in Figure 4.6c. However, at a higher frequency-product $f \cdot h = 500$ KHz $\cdot$ mm (Figure 4.6d), the curves of group velocities are not coincident. The results for the antisymmetric mode A_0 are also nearly equal at low values of the frequency-thickness product $f \cdot h = 11$ KHz $\cdot$ mm and considerably different at a much higher value of frequency-thickness product $f \cdot h = 500$ KHz $\cdot$ mm. These results are illustrating the fact that the plate theories give acceptable results only at low frequencies, while they fail for higher frequencies.

The composite plates considered before are symmetric with respect to their midplanes. However, the conclusions made about the frequency range of validity of the models based on plate theories are also true for non-symmetric plates. In Figure 4.7, the group wave surfaces of all fundamental wave modes qA_0, qSH_0 and qS_0 propagating in a cross-ply non-symmetric plate of CFRP-T700GC/M21 (Table A.1 in Ap-

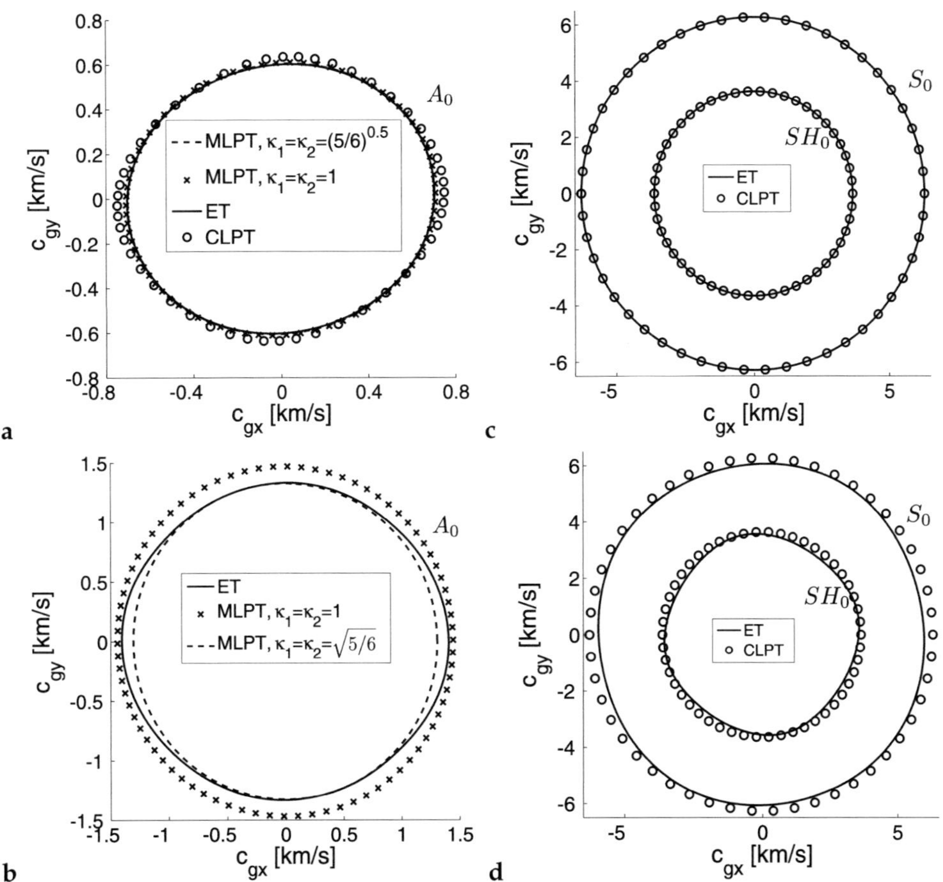

Figure 4.6: Group wave surfaces (GWS) of fundamental Lamb wave modes S_0, SH_0 and A_0 at frequency-thickness products $f \cdot h = 11$ KHz $\cdot$ mm (a and b) and $f \cdot h = 500$ KHz $\cdot$ mm (c and d) in a laminated plate of IM7-Cycom-977-3 with stacking sequence $[0/45/-45/90]_{2s}$

pendix A.9) with stacking sequence $[0/90/0/90]$ are plotted for frequency-thickness values $f \cdot h = 11$ KHz $\cdot$ mm (a) and $f \cdot h = 75$ KHz $\cdot$ mm (b) in dependence on φ. As before, the same notations are used for the curves obtained using CLPT, MLPT and elasticity theory. Analysis of Figure 4.7 shows that as well as for symmetric composites, at low frequencies the dispersion curves for non-symmetric composites calculated using different theories are almost fully coinciding, and with growth of frequency the difference between them increases. Thus, already at a frequency-thickness product of 75 KHz $\cdot$ mm, GWS of the quasi-antisymmetric wave mode calculated by means of models on the basis of CLPT and elasticity theory are considerably different. Note

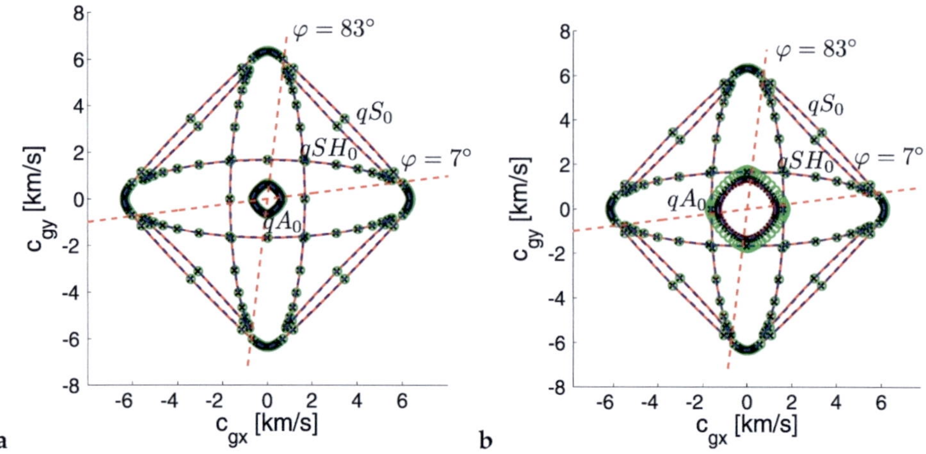

Figure 4.7: Group wave surfaces (GWS) of fundamental Lamb wave modes S_0, SH_0 and A_0 at frequency-thickness products $f \cdot h = 11$ KHz $\cdot$ mm (a) and $f \cdot h = 75$ KHz $\cdot$ mm (b) in a non-symmetric laminated plate of CFRP-T700GC/M21 with stacking sequence $[0/90/0/90]$. Values are calculated by solving the dispersion equation for CLPT: "'o'", MLPT ($\kappa_1 = \kappa_2 = 1$): "'x'", MLPT ($\kappa_1 = \kappa_2 = \sqrt{5/6}$): "'$--$'" and elasticity theory: "'$-$'"

that in contrast to the GWS of fundamental wave modes in quasi-isotropic laminates (Figures 4.5 and 4.6a and b), the influence of anisotropy on wave curves of fundamental wave modes in a cross-ply plate $[0/90/0/90]$ is much higher (Figure 4.7), i.e. the dependence of the dispersion properties on the propagation direction is stronger than for quasi-isotropic laminates (Figures 4.4, 4.5 and 4.6). Also the multifolding of the wave curve (multiple values of velocity) for the mode qSH_0 is observed in directions $\varphi \in [7°, 83°]$ (in the first quadrant). As it was described previously in section 2.3.5, the multiple (three) values of group velocities of wave front for the mode qSH_0 in one observation direction φ imply that the energy travels with this mode from the point source according to multiple (three) pulses at different group velocities. The directions of $\varphi = \varphi_{c1} = 7°$ and $\varphi = \varphi_{c2} = 83°$ are *caustics*[1] of the qSH_0 wave mode.

In Figure 4.8b and d the values of group velocities of wave fronts are presented for the symmetric plate $[0/90]_s$ (b) and non-symmetric $[0/90/0/90]$ (d) cross-ply plate for the frequency-thickness product $f \cdot h = 300$ KHz $\cdot$ mm. The values of wavenumbers $k_m(\gamma, \omega)$ are obtained by solving the corresponding dispersion equation obtained for the elasticity theory (4.1), then these values are used for plotting the slowness curves $s_m = k_m(\gamma, \omega)/\omega$ (Figure 4.8a and c) and applying the formula given by (2.55), wavenumbers are used for the calculation of the GWS (Figure 4.8b and d). In the

---

[1]The values $\varphi_{cj} + \pi/2$ are also the caustics due to the symmetry of wave curves.

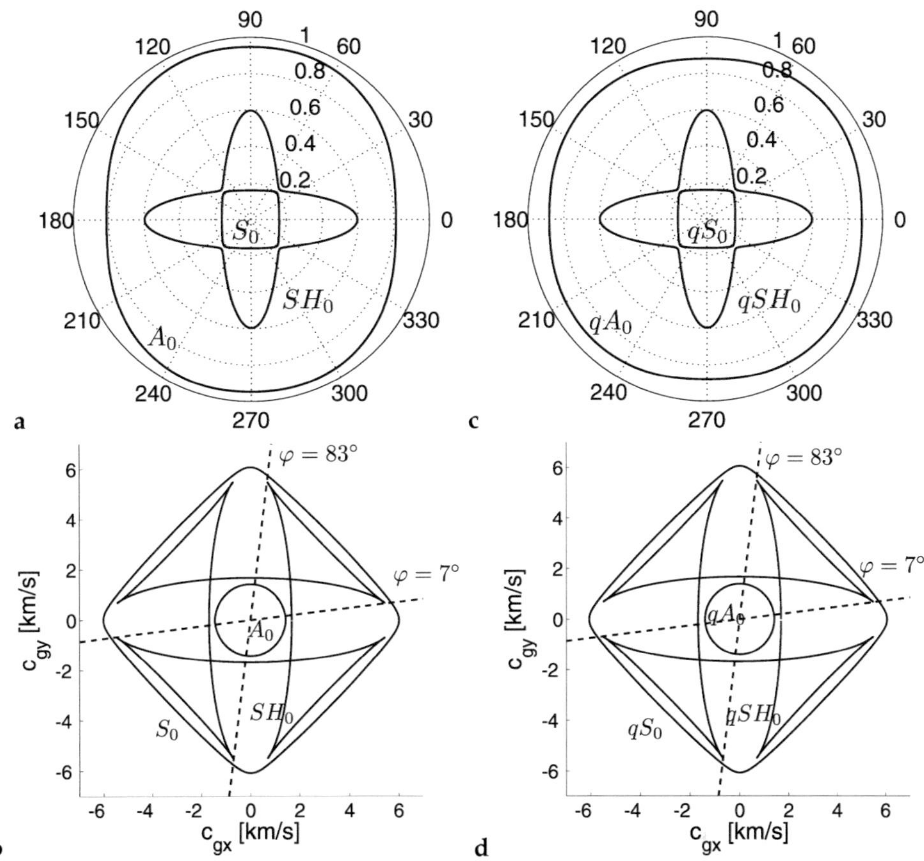

Figure 4.8: Slowness curves in s/km of fundamental Lamb modes in symmetric $[0/90]_s$ (a) and non-symmetric $[0/90/0/90]$ (c) composite plates made of CFRP-T700GC/M21 at frequency-thickness product 300 KHz · mm. Group velocities of wave fronts observed in $[0/90]_s$ (b) and $[0/90/0/90]$ (d) plates

case of the quasi-antisymmetric mode qA_0, the slowness curve is similar to that of the isotropic or quasi-isotropic materials. In constrast, the slownesses of the quasi-symmetric modes qS_0 and qSH_0 strongly depend on the propagation direction γ. A comparison of slowness curves for both plates leads to the conclusion that the swap of the last two layers in the plate does not influence the curves of quasi-symmetric and quasi-shear horizontal modes, but some changes for the qA_0 mode are visible, namely the dependence of slownesses on the angle γ is stronger for the symmetric plate $[0/90]_s$ than for the non-symmetric plate $[0/90/0/90]$. Due to the low dispersion of fundamental quasi-shear horizontal mode qSH_0 at frequencies below 500 KHz · mm and due to nearly the same values of slownesses for qSH_0 in both plates, the caustics are equal to the caustics φ_{c1} and φ_{c2} previously found for the non-symmetric plate at frequencies 11 and 75 KHz $\cdot$ mm, i.e. $\varphi_{c1} = 7°$ and $\varphi_{c2} = 83°$.

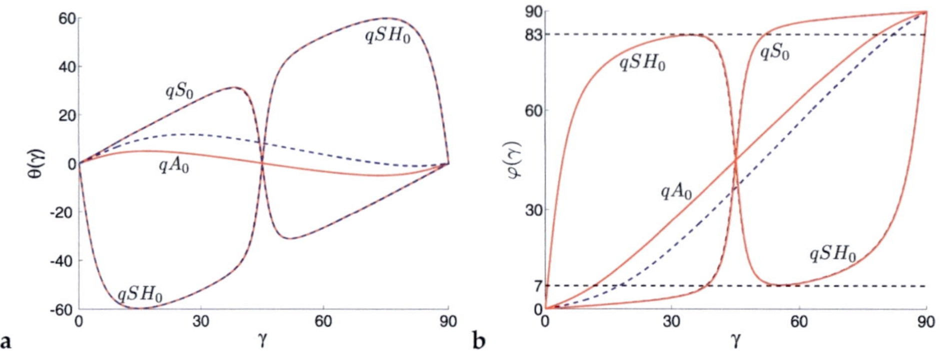

Figure 4.9: (a) Steering angle $\theta_m(\gamma)$ of Lamb wave modes in symmetric $[0/90]_s$ $(--)$ and non-symmetric $[0/90/0/90]$ $(-)$ composite plates at frequency-thickness product 300 KHz $\cdot$ mm. (b) Observation direction $\varphi_m(\gamma) = \gamma - \theta_m(\gamma)$ for wave excitation in direction γ

Note that the considerable difference between the slowness curves and group wave curves (surfaces) is explained by the fact that due the strong influence of the anisotropy on the wave modes in Figure 4.8, these waves strongly deviate from the incident angle γ, i.e. the values of deviation or steering angles $\theta(\gamma) = \gamma - \varphi$ (2.58) are high. In Figure 4.9a a steering angle $\theta(\gamma)$ in dependence on the propagation direction γ is shown for both $[0/90]_s$ and $[0/90/0/90]$ composite plates at the same frequency-thickness product $f \cdot h = 300$ KHz $\cdot$ mm. Its values are varying between $-60°$ and $60°$ for the quasi-shear horizontal mode qSH_0 and between $-30°$ and $30°$ for the quasi-symmetric mode qS_0. The maximum deviation of quasi-antisymmetric Lamb wave qA_0 from the propagation direction γ is $\pm 10°$ in the symmetric $[0/90]_s$ plate, while for the non-symmetric plate $[0/90/0/90]$ its value is about ± 4 degrees. Using the steering angle $\theta_m(\gamma)$, observation directions $\varphi(\gamma) = \gamma - \theta(\gamma)$ in dependence on incident angle γ are plotted (Figure 4.9b). This plot shows the directions in which Lamb waves with wave vector in the direction γ can be observed. As seen from Figure 4.9b, the functions $\varphi(\gamma)$ are not one-to-one for the quasi-shear horizontal mode qSH_0, and in observation directions between $7°$ and $83°$ more than one wave can be observed. It explains a previously detected multiplicity of group and phase velocities of Lamb waves.

The dispersion characteristics can also be plotted as surfaces in dependence on both angle φ (or γ) and frequency-thickness product $f \cdot h$ in 3D. The surfaces of group velocities of the wave front for S_0 (a) and SH_0 (c) Lamb modes are illustrated in Figure 4.10

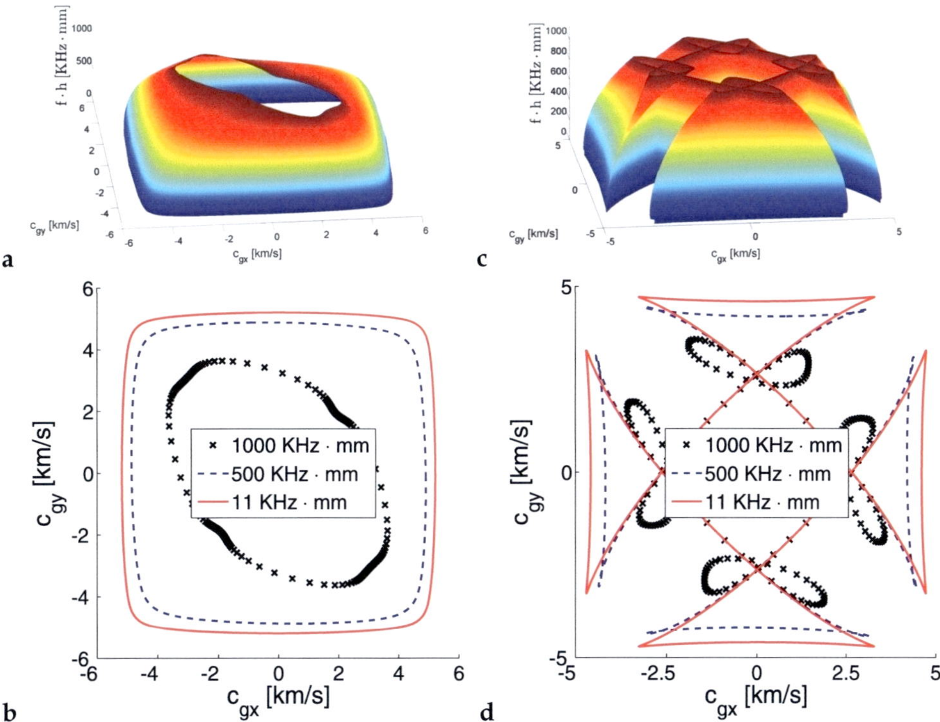

Figure 4.10: Group wave surfaces (GWS) of fundamental Lamb wave modes S_0 (a) and SH_0 (c) for frequencies lower than 1000 KHz · mm presented as a surface in dependence on φ (xy-plane) and $f \cdot h$ (vertical axis) in a cross-ply plate $[45_6/-45_6]_s$ made of CFRP-T700GC/M21. GWS for S_0 (b) and SH_0 (d) at fixed values of frequency-thickness product

for CFRP-T700GC/M21 (Table A.1 in Appendix A.9) plate with stacking sequence $[45_6/-45_6]_s$ in a frequency-thickness range $f \cdot h \leq 1000$ KHz · mm. The blue-colored and red-colored areas of these surfaces correspond to low and high frequencies, respectively. The group velocities of both modes S_0 and SH_0 in the low-frequency range are almost constant, i.e. the modes are low-dispersive. For higher frequencies the group velocity of both wave modes decreases and the form of their wave fronts changes. It is well displayed in Figures 4.10b and d, which represent the GWS curves of S_0 (b) and SH_0 (d) wave modes at three different values of frequency-thickness product 11 KHz · mm (straight lines), 500 KHz · mm (dashed lines) and 1000 KHz · mm ("x" markers). The dependence on angle for the fundamental symmetric wave modes becomes more complicated with increasing frequency. It is true even for the quasi-isotropic composites, as it can be concluded from the comparison of the group velocities of wave fronts of symmetric wave modes for a plate made of IM7-Cycom977-3

(Table A.1 in Appendix A.9) with stacking sequence $[0/45/-45/90]_{2s}$ computed at $f \cdot h = 1000$ KHz $\cdot$ mm (Figure 4.11a) with the GWS computed for lower frequencies $f \cdot h = 11$ KHz $\cdot$ mm (Figure 4.6c) and $f \cdot h = 500$ KHz $\cdot$ mm (Figure 4.6d).

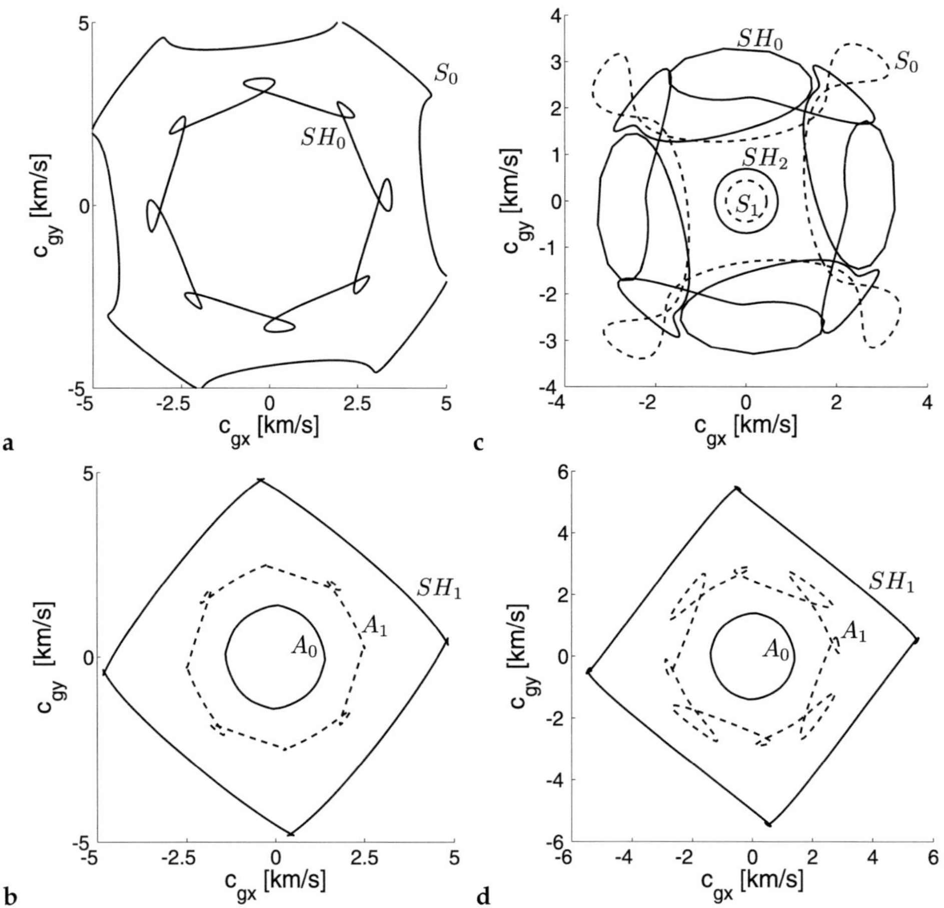

Figure 4.11: The GWS of symmetric (a and c) and antisymmetric (b and d) Lamb wave modes propagating at values of frequency-thickness product 1000 KHz $\cdot$ mm (a and b) and 1400 KHz $\cdot$ mm (c and d) in a quasi-isotropic plate $[0/45/-45/90]_{2s}$ made of IM7-Cycom977-3

The frequency-thickness product 1000 KHz $\cdot$ mm is higher than the cut-off frequencies[1] of higher-order antisymmetric wave modes A_1 (690 KHz $\cdot$ mm) and SH_1

---

[1] The cut-off frequencies of the first higher-order Lamb wave modes propagating in laminated composites considered in this thesis are listed in Table A.2 in Appendix A.9.

(700 KHz · mm), i.e. these wave modes are propagating at 1000 KHz · mm. The GWS of A_1, SH_1 and A_0 are plotted in Figure 4.11b. The dependence of the A_0 wave mode on the observation direction φ is weak, and with increasing frequency this dependence almost disappears, i.e. the wave front of this mode has a quasi-isotropic structure. However, the dependence of both higher-order modes A_1 and SH_1 on φ is complicated. These conclusions are valid for the dispersion curves of A_1, SH_1 and A_0 for the frequency-thickness product $f \cdot h = 1400$ KHz · mm (Figure 4.11d). At this value of $f \cdot h$, there are four propagating symmetric wave modes (Figure 4.11c), namely fundamental Lamb modes S_0 and SH_0 and higher-order Lamb modes S_1 and SH_2, which are observed at frequencies higher than their cut-off frequency[1] $f \cdot h = 1389$ KHz · mm. The curves of fundamental wave modes are complicated, whereas the curves of higher-order symmetric wave modes at the frequencies near to the cut-off frequency-thickness are quasi-isotropic.

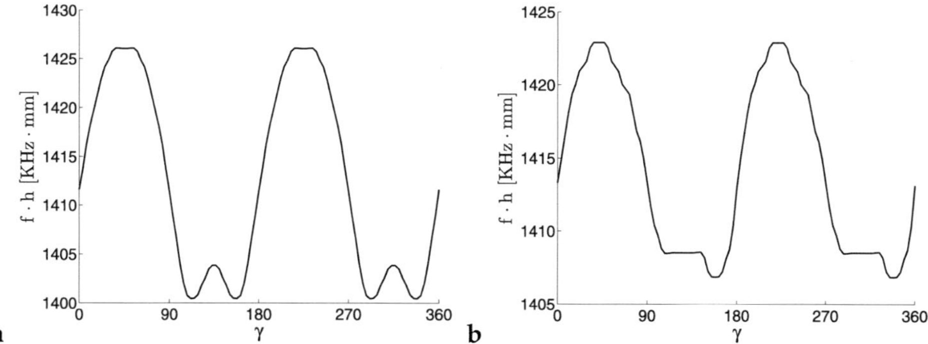

Figure 4.12: Frequency-thickness of first appearance of S_1 mode and backward mode S_{2b} in dependence on γ in two plates made of CFRP-T700GC/M21 with stacking sequences $[45_6/-45_6]_s$ (a) and $[45/-45/0/90]_s$ (b)

The Lamb wave modes studied in previous figures correspond to the regular poles, i.e. their group velocities are positive. Irregular poles first appear on the real axis only at high frequencies. Moreover, for some frequency ranges, irregular poles (in contrast to regular poles) approach the real axis not simultaneously. For example, this phenomenon is observed for values of frequency-thickness product between 1400 and 1426 KHz · mm in a symmetric composite plate made of AS4/3502 with stacking sequence $[45_6/-45_6]_s$. The Lamb wave mode corresponding to this irregular pole is named S_{2b} ("b" stands for backward mode). At the same frequencies, the wave mode S_1 corresponding to a regular pole also becomes to be propagating. Figure 4.12a illustrates the dependence of the frequency-thickness of the first appearance of the back-

---

[1]These modes are decoupled at this frequency, i.e. they correspond to poles of different components of Green's matrix.

ward mode S_{2b} (as well as S_1 mode) on the propagation direction γ. In Figure 4.12b the frequency-thickness of first appearance of the S_{2b} wave mode in a quasi-isotropic plate $[45/-45/0/90]_s$ made of AS4/3502 is plotted. The wave modes S_1 and S_{2b} in a frequency range between minimal and maximal frequency-thickness of first appearance are propagating not in all directions.

The wavenumber curves of symmetric wave modes propagating in a frequency-thickness range $f \cdot h \in [1300, 1430]$ (in KHz $\cdot$ mm) are presented for the AS4/3502 laminated plate with stacking sequence $[45_6/-45_6]_s$ in Figures 4.14a and b for directions $\gamma = 0°$ and $\gamma = 45°$, respectively. The wave modes S_1 and S_{2b} appear firstly at different frequencies and with different values of wavenumbers. The dependence of wavenumbers for both wave modes on γ is shown for this composite plate in Figures 4.13a, b and c, when the roots of dispersion equation are calculated for the frequency-thickness products 1425 KHz $\cdot$ mm (a), 1400 KHz $\cdot$ mm (b) and 1390 KHz $\cdot$ mm (c). In Figures 4.13d, e and f, the corresponding values of attenuation coefficients for this mode are plotted in directions γ, in which both wave modes are non-propagating. These directions are clearly observed in Figures 4.13d, e and f, while the real parts of wavenumbers of both modes are equal in these directions. Note that at $f \cdot h = 1390$ KHz $\cdot$ mm, both modes S_1 and S_{2b} are non-propagating, the values of wavenumbers of S_{2b} have the same real part as the wavenumbers of S_1 but their imaginary parts are of opposite sign, however both modes are represented by the same curves in Figure 4.13f.

Remark 4.4 *The algorithm for the evaluation of Green's matrix described in chapter 3 can be applied for modelling of the wave propagation in a composite plate with clamped lower boundary. It can be done by considering the matrix*

$$\mathbf{S}^{(N)}_{\text{clamped}} = \begin{pmatrix} 1 & 0 & 0 & 0 & 0 & 0 \\ 0 & 1 & 0 & 0 & 0 & 0 \\ 0 & 0 & 1 & 0 & 0 & 0 \end{pmatrix} \tag{4.16}$$

instead of the matrix $\mathbf{S}^{(N)}$ in (3.39). An example of dispersion curves for a composite plate with a clamped lower boundary is given in Figure 4.15 as wavenumber curves in dependence on γ at frequency-product values 500 KHz $\cdot$ mm (a) and 710 KHz $\cdot$ mm (b). The corresponding wave modes are surface waves of the Rayleigh wave type.

As shown in this section, Lamb waves in multilayered composite plates can have a very complex structure, and therefore an analysis of dispersion properties is an important step in the development of methods for structural health monitoring [20].

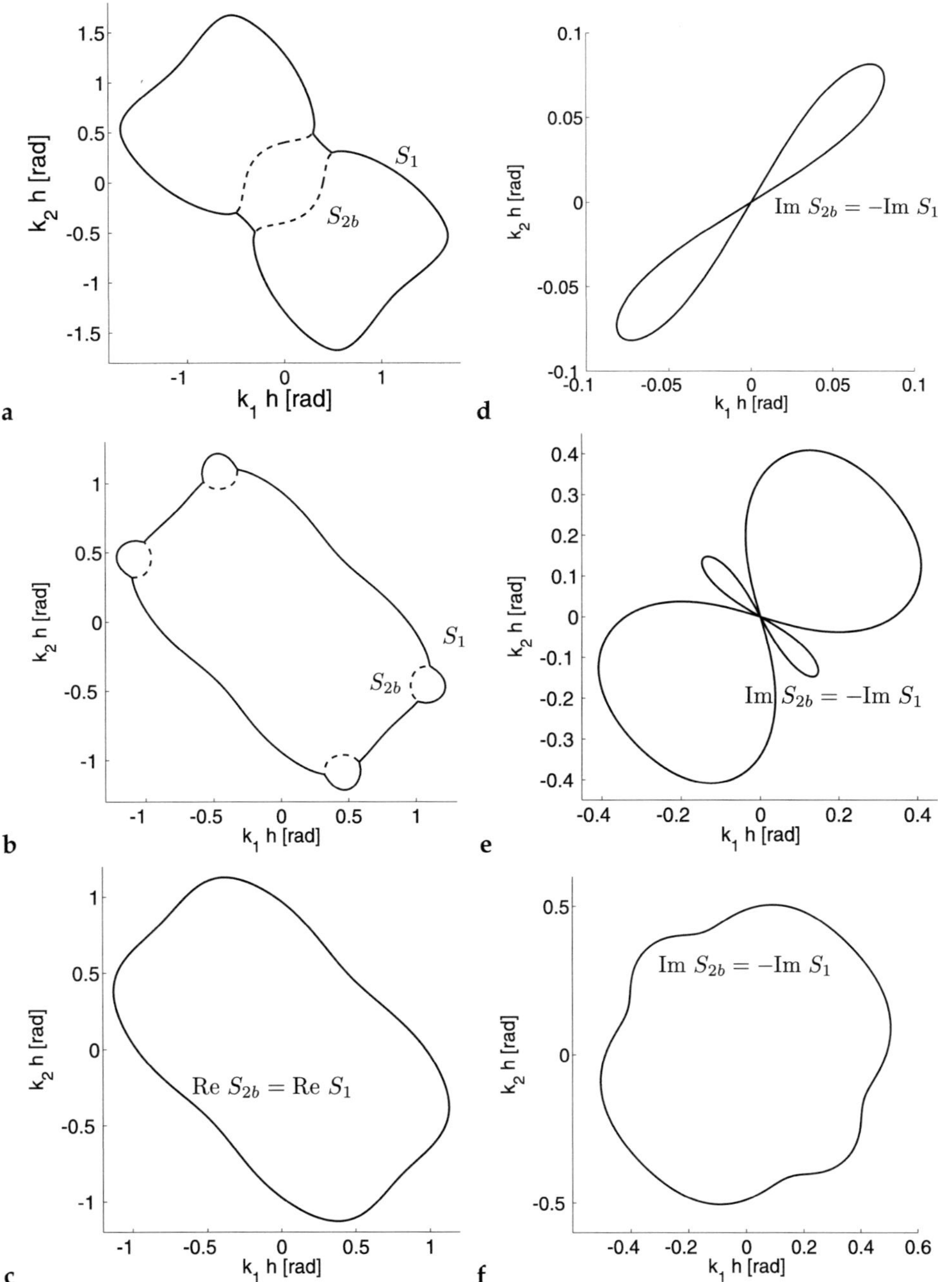

Figure 4.13: Real (a, b and c) and imaginary (d, e and f) parts of wavenumbers of S_1 and S_{2b} wave modes in dependence on γ in a laminated plate $[45_6/-45_6]_s$ made of AS4/3502 at frequency-thickness products 1425 KHz · mm (a and d), 1400 KHz · mm (b and e) and 1390 KHz · mm (c and f)

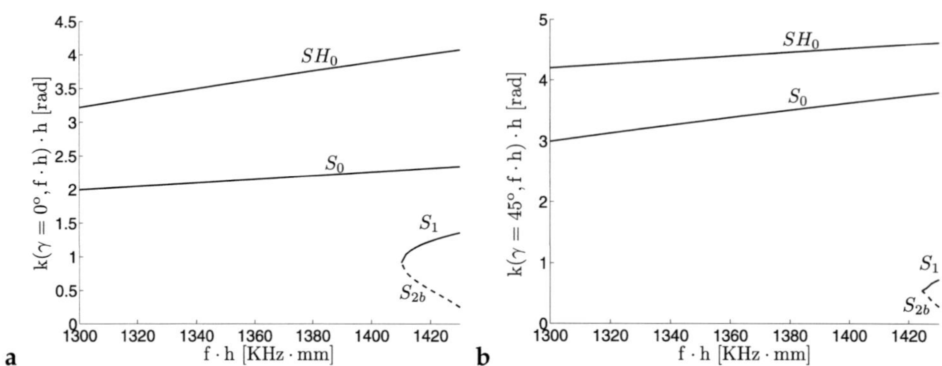

Figure 4.14: Wavenumbers of symmetric wave modes propagating at frequencies $f \cdot h$ $\in [1300 \text{ KHz} \cdot \text{mm}, 1430 \text{ KHz} \cdot \text{mm}]$ in a laminated plate of AS4/3502 with stacking sequence $[45_6/-45_6]_s$, plotted in directions $\gamma = 0°$ (a) and $\gamma = 45°$ (b)

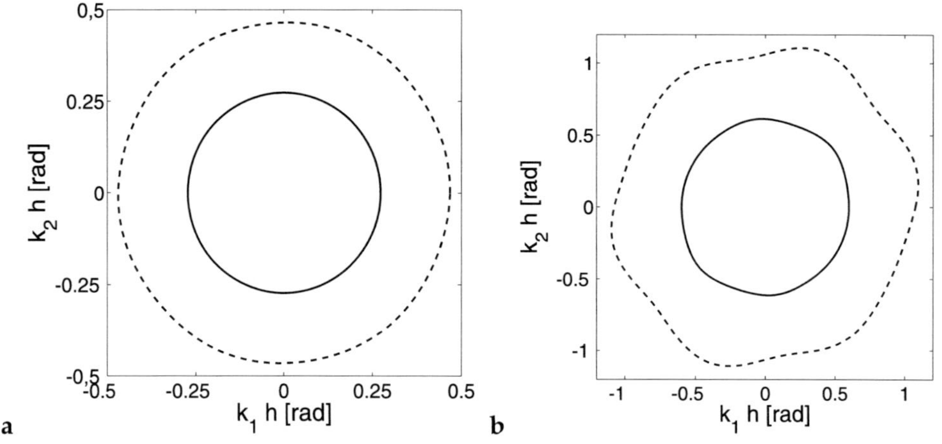

Figure 4.15: Wavenumbers of propagating wave modes in a plate made of AS4/3502 with stacking sequence $[45/-45/0/90]_s$ with lower boundary clamped. Curves are calculated in dependence on γ at fixed frequency-product values 500 KHz $\cdot$ mm (a) and 710 KHz $\cdot$ mm (b)

5 Methods of evaluation of two-dimensional wavenumber integral

As expressed in chapter 3, the application of an integral approach based on the Fourier transform after finding the solution of the problem in Fourier domain requires the computation of the inverse Fourier transform (3.10), i.e. the computation of two-dimensional wavenumber contour integral and consecutive evaluation of the integral with respect to the frequency ω. The most time-consuming step is the computation of the double integral over wavenumbers

$$\mathbf{u}(r,\varphi,z,\omega) = \frac{1}{4\pi^2} \int\limits_{0}^{2\pi} \int\limits_{\Gamma^+(\gamma)} \mathbf{K}(\alpha,\gamma,z,\omega)\mathbf{Q}(\alpha,\gamma)\,e^{-i\alpha r \cos(\gamma-\varphi)}\alpha\,d\alpha d\gamma, \qquad (5.1)$$

which causes difficulties such as integral singularity near real poles (or branch points of Green's matrix, if they present, e.g. for a half-space) of Green's matrix, strong oscillations of the integrand and significant time expenses. In this chapter, the methods of the evaluation of the double integral (5.1) pare given in detail and compared by numerical examples. Note that some parts of this research are presented by the author of this thesis at several conferences [57, 58, 59, 60] and published in international journals [58, 62].

5.1 Short overview on methods of evaluation of $2D$-wavenumber integral

Methods of evaluation of the $2D$-wavenumber integral (5.1) are differing depending on the spatial domain with respect to the excitation source, in which their application is more effective/more precise. It is known that at short distances from the excitation source, the wave field is garbled by the interaction of multiple waves from all parts of the source surface. This is confined to the region called the *near-field*

$$r \leq A_o^2/\lambda_{\min} = A_o^2 k_{\max}/(2\pi), \qquad (5.2)$$

where A_o corresponds to the radius of the circle which contains the loading domain Ω and λ ($k_{\max}$) is a minimal wavelength (maximal wavenumber) of the waves. The far-field is the region outside the near-field where the transducer waves coalesce to produce a plane wave whose on-axis intensity decreases inversely with distance [75]. Note that according to (5.2), a large excitation source produces a large near-field and is thus not suitable for use on a thin object [75].

The evaluation of the $2D$-wavenumber integral (5.1) is difficult due to the presence of singularities within the integration domain and the highly oscillatory nature of the integrands at higher frequencies and large distances between the field and source points. Moreover, to date there is no algorithm available to evaluate the double wavenumber integrals [33]. In the near-field to an excitation source, contour integrals of the inverse Fourier transform (5.1) can be evaluated directly using adaptive two-dimensional numerical integration schemes (Clenshaw-Curtis quadrature schemes) [84, 140, 153]. Another approach to the integration of slowly decreasing oscillating functions is suggested in [123]. But the computational costs are still considerable since these approaches require millions of function evaluations. Another disadvantage lies in the fact that the wave structure of the solution using these techniques is not taken into account and the wave field cannot be analysed for each Lamb wave mode separately.

Other methods for the computation of the double wavenumber integral are based on the residue theorem or on the modal expansion technique, which allow to reduce computational costs considerably and to analyse the contribution of each Lamb wave mode. Thus, Lamb waves excited by point loads in isotropic [12] and anisotropic [26] plates are obtained using the hybrid numerical method and modal expansion technique. The steady-state vibrations are evaluated in the wavenumber domain by means of double sums: the inner sum with respect to normal modes and the outer sum with respect to the propagation directions of plane waves. Wave propagation from surface-bonded piezoeletric actuators in an isotropic plate is considered in [108, 119] using the residue theorem. A similar approach is originally offered for isotropic media in case of an axisymmetric loading [10]. When computing a wave propagation from surface-bonded piezo-actuators in an anisotropic layered plate, the authors in [107] substituted the integration along the positive real semi-axis by an integration along the whole real axis. The integral along the real axis is evaluated as before using the residue theorem. The integral over the angle γ of wavenumbers is then evaluated using the stationary phase method. In [13], the double integral over wavenumbers obtained for CLPT-based model is considered in Cartesian coordinates and is reduced by the use of the residue theorem to a one-dimensional integral taking into account the analytical representation of dispersion curves. The remaining improper integral is then computed numerically using standard quadratures. In [141], a far-field asymptotic expansion of related two-dimensional wavenumber integrals is obtained again using the stationary phase method in terms of the modal solutions to the forced 2D problem (i.e. using excitability matrices) for the case of loading by point sources. In a research carried out by the group of E. Glushkov [40, 43, 69] at the same time as this work was performed,

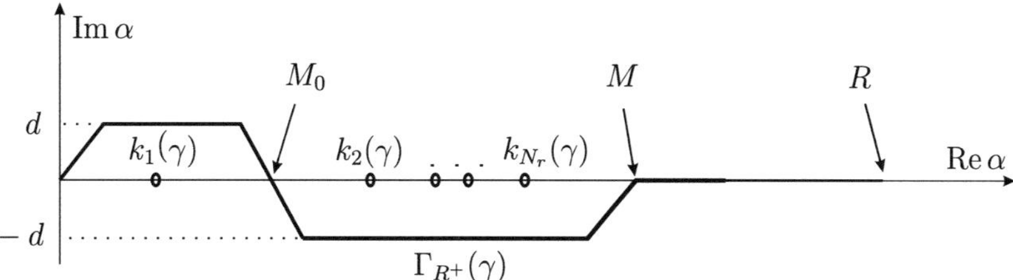

Figure 5.1: Finite integration contour $\Gamma_{R^+}(\gamma)$ in case all real poles except of only one irregular pole $k_1(\gamma)$ are regular; $d = \operatorname{Im} k_{nc}(\gamma)/2$ is the value of the deviation of the integration contour from the real axis into the complex plane, $M = \max\limits_{m} k_m(\gamma)$ and R is such a value that $M < \operatorname{Re}\alpha \leq R$

the results of Velichko and Wilcox [141] are extended to the case of the general surface load. The approximation of order $\mathcal{O}(r^{-1})$, $r = \sqrt{x^2 + y^2}$ for the computation of the integral (5.1) in a far-field to the excitation source is obtained applying again the stationary phase method.

In spite of many publications referenced here, wave propagation from common types of sources of finite size anisotropic layered composites has not yet been completely analysed. As it will be shown in this chapter, the integral (5.1) can be evaluated in a far-field with an error not more than $\mathcal{O}(r^{-2})$. Moreover, it will be shown on numerical examples that the formulas obtained in this thesis give good results under certain conditions for the analysis in the near-field to an excitation source. Additionally, the asymptotic expansion used by many authors [40, 43, 69, 107, 109, 141] is extended in this thesis to the case of the calculation of displacements near caustics. Finally, the results of different techniques described in this thesis are compared to each other, the pros and contras as well as the limits of the applicability of each technique are discussed.

5.2 Method of direct integration using adaptive quadratures

In this section the direct method of 2D-wavenumber integral computation using adaptive quadratures is discussed. This approach is generally similar to the adaptive quadratures used in [84, 153] and can be applied for the computation of displacements based on the elasticity (2.5) as well as on plate theories (2.29), (2.36).

5.2.1 The convergence of $2D$-wavenumber integral

If the double integral (5.1) is considered for $z < 0$, it is obtained that it converges at all points of xy-plane since, if $\alpha \to \infty$, the components of Green's matrix[1] contain decaying exponents, Equations (3.55), (3.56), and the function under the double integral sign is absolutely integrable [10]. Moreover, according to Fubini's theorem it can be calculated as an iterated integral [118]. For $z = 0$, the convergence of the double integral according to (5.1) depends on the properties of the load vector $\mathbf{Q}(\alpha, \gamma)$ in transformed domain:

$$\text{if } \mathbf{Q}(\alpha, \gamma) \sim \mathrm{o}(\alpha^{-1}), \quad \alpha \to \infty, \tag{5.3}$$

then the function in the integral is absolutely integrable for all r and φ. If the condition (5.3) is not satisfied, then the function in the integral sign is not absolutely integrable, however the integral converges due to the oscillations of exponential term $\exp(-i\alpha r \cos(\gamma - \varphi))$ for all values of r and φ except of origin $r = 0$, where a logarithmic singularity occurs [10, 145]:

$$\mathbf{u}(r, \varphi, 0, \omega) = \mathbf{u}_0(\varphi, 0, \omega) \log r + \mathbf{u}_1(r, \varphi, 0, \omega), \quad r \to 0, \tag{5.4}$$

where $\mathbf{u}_0(\varphi, 0, \omega)$ and $\mathbf{u}_1(r, \varphi, 0, \omega)$ are continuous functions of their variables.

Note that if the double integral (5.1) is not converging absolutely for $z = 0, r > 0$, the corresponding iterated integrals can be diverging or lead to the results, different from the value of original double integral [118]. Nevertheless, instead of the improper double integral (5.1) the sequence of following double integrals can be considered:

$$\mathbf{u}_{R_n}(r, \varphi, z, \omega) = \frac{1}{4\pi^2} \int\limits_0^{2\pi} \int\limits_{\Gamma_{R_n}^+(\gamma)} \mathbf{K}(\alpha, \gamma, z, \omega) \mathbf{Q}(\alpha, \gamma) \, e^{-i\alpha r \cos(\gamma - \varphi)} \alpha \, d\alpha d\gamma, \tag{5.5}$$

where $\Gamma_{R_n}^+ \to \Gamma^+$, $n \to \infty$. Obviously, this sequence converges to the initial $2D$-wavenumber integral, i.e. $\forall \varepsilon > 0, \forall \omega, \varphi, z, \forall r > 0 \ \exists n > 0$ that

$$\left| u_j(r, \varphi, z, \omega) - u_{j,R_n}(r, \varphi, z, \omega) \right| < \varepsilon, \quad j = 1, 2, 3. \tag{5.6}$$

It is concluded that instead of the improper double integral (5.1) the double integral over the bounded domain (5.5) can be considered, i.e. for given value of ε such $R = R_n$ exists that

$$\mathbf{u}(r, \varphi, z, \omega) \approx \mathbf{u}_R(r, \varphi, z, \omega), \tag{5.7}$$

---

[1] All procedures of the wavenumber integral evaluation given in this chapter can be in the same way applied for models based on CLPT and MLPT. If needed, the differences in the application are explicitly discussed.

and (5.6) is satisfied. Such a contour $\Gamma_R^+(\gamma)$, in the case that all real poles except of only one irregular pole $k_1(\gamma)$ are regular, is presented in Figure 5.1. Note that since double integral (5.7) converges absolutely, it can be evaluated as an iterated integral [118]. In the following the evaluation of the 2D-wavenumber integral (5.1) as an iterated integral (5.7) is called *Direct Contour Integration* (DCI).

Remark 5.1 *Due to the asymptotic properties of Green's matrices for models based on CLPT (3.77) and MLPT (3.75), the inner integrals with respect to α in (5.1) are converging[1] even for a concentrated point source, see Equation (3.17). The speed of convergence of (5.7) with an increasing of R is due to the asymptotic properties of corresponding Green's matrices (3.78) higher than in case of modelling by elasticity theory.*

5.2.2 Estimation of the truncation error in case of excitation by point source

In this section the truncation error of formula (5.7) with respect to the exact formula (5.1) is analysed for the case of wave excitation by point source in direction x_j, $j = 1,2,3$ (see Equation (3.17)). The difference between the exact and the approximate expressions for displacements is given for the i-th component of displacement vector as

$$u_i(r,\varphi,z,\omega) - u_{i,R}(r,\varphi,z,\omega) = \frac{1}{4\pi^2} \int\limits_0^{2\pi} \int\limits_R^\infty K_{ij}(\alpha,\gamma,z,\omega)\,e^{-i\alpha r\cos(\gamma-\varphi)}\alpha\,d\alpha d\gamma, \quad (5.8)$$

where instead of Green's matrix its asymptotics for a model based on the elasticity theory (3.70) can be applied for $\alpha > R \gg 1$:

$$\frac{1}{4\pi^2} \int\limits_0^{2\pi} \int\limits_R^\infty K_{ij}(\alpha,\gamma,z)\,e^{-i\alpha r\cos(\gamma-\varphi)}\alpha\,d\alpha d\gamma \qquad (5.9)$$

$$\approx \frac{1}{4\pi^2} \int\limits_R^\infty \alpha \int\limits_0^{2\pi} T_{ij}(\gamma,z,\omega)\,e^{-i\alpha r\cos(\gamma-\varphi)}\,d\gamma d\alpha + o(1).$$

Then, the main term in (5.9) can be estimated by applying the stationary phase method to the integral with respect to γ in (5.9) (for detail on stationary phase method see Appendix B.2): as $\alpha > R \gg 1$ and $r > 1$, the value $\alpha r \gg 1$ and it follows that

$$\frac{1}{4\pi^2} \int\limits_R^\infty \int\limits_0^{2\pi} T_{ij}(\gamma,z,\omega)\,e^{-i\alpha r\cos(\gamma-\varphi)}\,d\gamma d\alpha = \sqrt{\frac{1}{8\pi^3}}\left(T_{ij}(\varphi,z,\omega)\,e^{i\pi/4} \int\limits_R^\infty \sqrt{\frac{1}{r\alpha}} \right.$$

$$\left. \times \quad e^{-i\alpha r}\,d\alpha + T_{ij}(\varphi+\pi,z,\omega)\,e^{-i\pi/4} \int\limits_R^\infty \sqrt{\frac{1}{r\alpha}}\,e^{i\alpha r}\,d\alpha \right), \qquad (5.10)$$

---

[1]Except of values $\gamma = \varphi + \pi/2$, $\gamma = \varphi + 3\pi/2$, where due to the absence of an oscillation the logarithmic singularity can occur, which is integrable when evaluating the integral with respect to γ.

for $i, j = 1, 2, 3$, and where stationary points are found to be $\gamma_1 = \varphi$, $\gamma_2 = \varphi + \pi$. The improper integrals with respect to α in (5.10) converge due to the oscillation of the exponent and due to the decaying multiplier $1/\sqrt{\alpha}$. Using an integration by parts for the integrals with respect to α in (5.10), it is obtained

$$\int_R^\infty \sqrt{\frac{1}{r\alpha}}\, e^{\pm i\alpha r}\, d\alpha = \pm \frac{i}{r\sqrt{rR}}\, e^{\pm iRr} + \mathcal{O}(r^{-5/2}R^{-3/2}), \tag{5.11}$$

i.e. the residual (5.8) tends to zero, because $r^{-3/2}R^{-1/2}$ tends to zero with increasing values r or R. This means that the truncation of the contour $\Gamma^+(\gamma)$ is valid in a near-field for high enough values of R and in a far-field where is no need to consider such high values of R and the computation time can be saved.

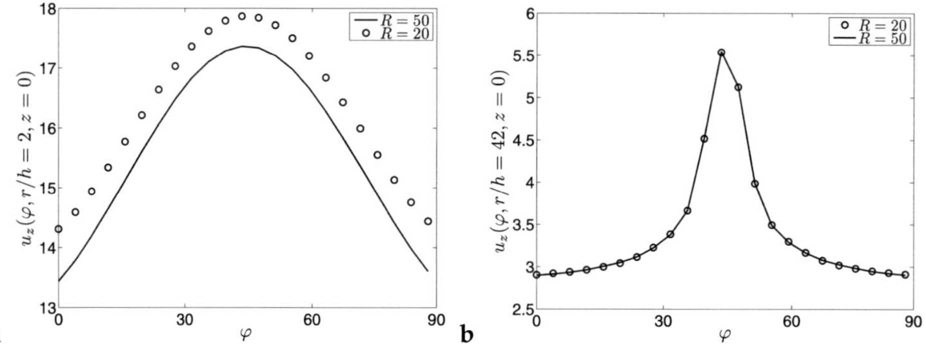

Figure 5.2: $|u_z(\varphi)|$ calculated by DCI (Equation (5.7)) when $f \cdot h = 500$ KHz $\cdot$ mm, $r/h = 2$ (a) and $r/h = 42$ (b), $z = 0$ for $R = 50$ and for $R = 20$ for an excitation source given by a concentrated vertical force at $r = 0$ on the surface of the cross-ply $[45_6/ - 45_6]_s$ plate made of AS4/3502 (Table A.1)

The dependence on r and R is illustrated in Figure 5.2 where the results of the computation of $|u_z(\varphi)|$ at values $r/h = 2$ (a) and $r/h = 42$ (b) for the values of $R = 20$ ("$-$") and $R = 50$ ("o") are compared. Whereas at $r/h = 2$ the amplitudes of $u_z(\varphi)$ are similar but not equal for $R = 20$ and $R = 50$, the difference between the corresponding results seem to be neglible at $r/h = 42$.

Remark 5.2 *Since the loading functions can be approximated by the sum of the point sources (see section 3.1.4.5), for the load of general type the corresponding error estimation can be obtained using formula (5.11). Moreover, the wavenumber representation of distributed sources $\mathbf{Q}(\alpha, \gamma)$, e.g. for representations given in section 3.1.4, the decay rate of the estimation error with respect to R is usually higher than in case of the point excitation sources, see Equation (5.11).*

5.2.3 Details on the computation of wavenumber integral

Applying the representation (5.7) for the computation of displacements, it is needed to choose the contour $\Gamma_R^+(\gamma)$ according to the principle of limiting absorption (see section 3.1.3) as deviating from the positive real semi-axis while enclosing real poles of Green's matrix in the complex plane α. However, due to the presence of only a finite number of real poles, the value $M = \max_m k_m(\gamma)$ exists, where $k_m(\gamma)$, $m = 1, \ldots, N_r$ are real poles. For values $\mathrm{Re}\,\alpha > M$ the contour $\Gamma^+(\gamma)$ coincides with the real axis. For numerical calculation, the contour $\Gamma_R^+(\gamma)$ should be chosen to be far enough from the real axis for values $\mathrm{Re}\,\alpha \approx k_m(\gamma)$, $m = 1, \ldots, N_r$, in order to avoid the high gradient of Green's matrix components near to real singularities. However, the complex poles with negative imaginary part should be enclosed by the contour from above. The optimal value of the deviation d of the integration contour from the real axis into the complex plane should be $d = \mathrm{Im}\,k_{nc}(\gamma)/2$, where $k_{nc}(\gamma)$ is the complex pole nearest to the real axis[1].

Remark 5.3 *Instead of the real-valued frequency $\omega = \omega_R$, the complex frequency $\omega_C = \omega_R + i\omega_I$ where $0 < \omega_I \ll \omega_R$, can be considered. Then, the integral with respect to α in (5.7) is evaluated directly along the real semi-axis since all real poles shift from the real axis into the complex plane. Note that introducing the complex frequency is equivalent to introducing of an internal friction, i.e. the waves become to be attenuating and the inaccuracy of such a modelling in comparison to the exact elastic model grows as r increases [58]. It requires the choice of small ω_I which in turn, requires a large number of grid spacings near to the poles appearing close to the real axis.*

However, it seems to be the only one possible approach in case the frequency ω equals to the resonance frequency since according to the principle of limiting absorption the wavenumber integral cannot be evaluated as standard definite integral (see section 3.1.3).

Then, the double integral (5.7) can be evaluated as an iterated integral with respect to α and γ. However, integrals with respect to both variables are oscillating with respect to r, i.e. in the far-field, with an increasing of r, the computational time needed for the computation of the integral (5.7) increases significantly comparing to the time needed for its evaluation at points in the near-field of the excitation source. It is especially appreciable due to the considerable costs of the computation of Green's matrix for each parameter pair (α, γ) in case of the high number of layers in a plate. However, for each z and ω, the components of Green's matrix are smooth functions of γ and α if α lies on the contour $\Gamma_R^+(\gamma)$. Green's matrix components in dependence on γ and α can be interpolated on a two-dimensional grid[2], and then, these interpolations will be used instead of the corresponding Green's matrix components. Finally, the oscillating

---

[1] In case of Green's matrix for the models based on CLPT and MLPT without taking into account internal friction, there are no complex poles and the value d can be chosen arbitrarily, e.g. $d = 0.2$.

[2] It is prefered to use a non-regular grid if the integration contour goes near to the real or complex poles as it produces the high gradient of some Green's matrix components near these poles.

integrals with respect to both variables can be evaluated using some adaptive quadratures, such as used in [123, 140, 153] or the commercial Fortran subroutine D01AKF of NAG Library [1].

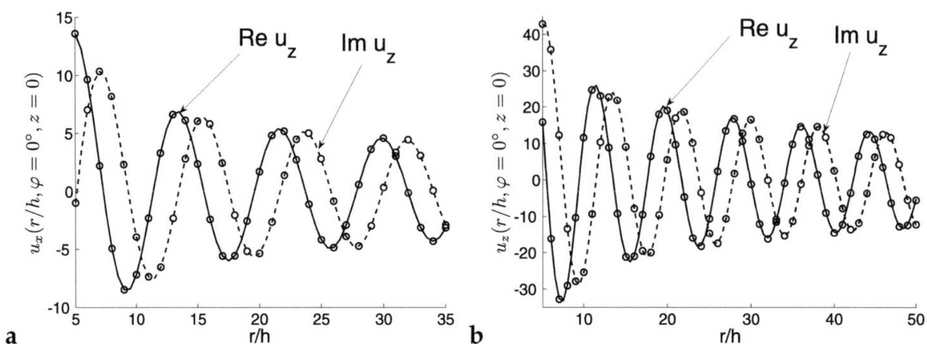

Figure 5.3: Results of the calculation of the double integral (5.7) for complex amplitudes of $u_x(r/h)$ (a) and $u_z(r/h)$ (b) when $f \cdot h = 100$ KHz $\cdot$ mm, $\varphi = 0°$, $z = 0$: the straight (Re) and the dotted lines (Im) correspond to $R = 10$, circles correspond to $R = 500$. An excitation source is given by a concentrated vertical force at $r = 0$ on the surface of a cross-ply $[45_6/ - 45_6]_s$ plate made of AS4/3502 (Table A.1). Plate is modelled applying MLPT $(\kappa_1 = \kappa_2 = \sqrt{5/6})$

An example of the calculation of the wavenumber integral (5.7) for the model based on the MLPT $(\kappa_1 = \kappa_2 = \sqrt{5/6})$ is illustrated in Figure 5.3 for different values of r in a direction $\varphi = 0°$ on the surface $z = 0$. The harmonic wave propagation is excited by a point vertical force at $r = 0$, see Equation (3.17), for the frequency-thickness product $f \cdot h = 100$ KHz $\cdot$ mm. As it can be seen, the results for $R = 10$ and $R = 500$ are equal to each other already at distances about $r/h < 10$, so the use of the value $R = 10$ is sufficient for a good accuracy in a computation of the double integral (5.7) for Green's matrix based on the MLPT. However, in case of the models based on the elasticity theory, the recommended value for R is about $R = 50$ and depends on the required accuracy of the displacement vector.

The algorithm described in this section reduces the amount of calculation for the evaluation of (5.1) considerably and can be applied for studying the harmonic wave propagation in anisotropic laminates excited by surface source. However, the computational resources needed for the evaluation of wavenumber integral in the far-field are significant and the analysis of a transient problem is still very time consuming since the integral (5.1) needs to be evaluated for a wide range of frequencies. Moreover, the presented algorithm does not take into account the wave structure of the solution, i.e. the wave modes cannot be studied separately of each other.

5.3 Cauchy's residue theorem-based method for evaluation of the 2D-wavenumber integral

In this section an algorithm of the computation of the wavenumber integral (5.1) based on Cauchy's residue theorem for the computation of the improper contour integral with respect to α and consequent numerical integration with respect to γ is presented. Similar techniques are known for anisotropic layered structures for some cases of point loads [20, 141], circular and square piezo-electrical actuators [107, 109]. The procedure presented here extends this technique to the case of a load of general type, which can be represented in the wavenumber domain in the form

$$\mathbf{Q}(\alpha, \gamma) = \sum_j e^{-i\alpha g_j(\gamma)} \widetilde{\mathbf{Q}}_j(\alpha, \gamma), \tag{5.12}$$

where the functions $\widetilde{\mathbf{Q}}_j(\alpha, \gamma)$ and $g_j(\gamma)$ are known. The functions $\widetilde{\mathbf{Q}}_j(\alpha, \gamma)$ should be at least bounded functions for $\operatorname{Im}\alpha \to +\infty$ or $\operatorname{Im}\alpha \to -\infty$. Note that such a representation in a wavenumber domain can be obtained for each load function, at least approximately, since the load function can be approximated by the sum of point sources with some coefficients (see Equation (3.25)), leading to the terms

$$\mathbf{Q}(\alpha, \gamma) \approx \delta_s^2 \sum_{j=1}^{N_q} \mathbf{q}_j\, e^{i\alpha r_j \cos(\gamma - \varphi_j)}, \tag{5.13}$$

corresponding to point loads at $(x_j, y_j) = (r_j \cos \varphi_j, r_j \sin \varphi_j)$, $j = 1, \ldots, N_q$. In notations of (5.12), the following coefficients can be obtained: $q_j \equiv \widetilde{\mathbf{Q}}_j(\alpha, \gamma)$ and $g_j(\gamma) = -r_j \cos(\gamma - \varphi_j)$.

However, note that such a representation in a wavenumber domain can often be obtained also for the analytical wavenumber domain representation of the surface load. For example, the Bessel function $J_1(A_o\alpha)$ in the load function corresponding to the circular PZT disk of a radius A_o given by (3.17) using the property of Bessel function (Equation (B.15) in Appendix B.3) can be splitted into two terms as follows

$$
\begin{aligned}
J_1(A_o\alpha) &= \frac{H_1^{(1)}(A_o\alpha) + H_1^{(2)}(A_o\alpha)}{2} \tag{5.14} \\
&= \frac{e^{-i\alpha(-A_o)}\widetilde{H}_1^{(1)}(A_o\alpha) + e^{-i\alpha A_o}\widetilde{H}_1^{(2)}(A_o\alpha)}{2},
\end{aligned}
$$

where $H_1^{(j)}(A_o\alpha)$ is a Hankel function of kind j and of first order and $\widetilde{H}_1^{(j)}(A_o\alpha)$ is the same function scaled by the factor $\exp((-1)^j i\alpha A_o)$.

Similarly, the wavenumber domain representation of the loads frequently used in practical applications (see section 3.1.4) can be brought to the form (5.12) and the computational algorithm for evaluation of wavenumber integrals for loads of this form is of a great interest for the investigation of propagating Lamb waves.

Remark 5.4 *In the case of the vertical point load located at the origin, its wavenumber representation is according to (3.17) a constant, i.e. in the form (5.12) $g_j(\gamma) \equiv 0$:*

$$Q_3(\alpha, \gamma) = e^{-i\alpha \cdot g_1(\gamma)} \equiv 1. \tag{5.15}$$

5.3.1 Algorithm of calculation of the $2D$-wavenumber integral with respect to α in case of a load of general type

Assuming that a surface excitation source can be represented in the form (5.12), the displacement vector (5.1) can be calculated for each term j in the representation of load (5.12) as

$$\mathbf{u}(r, \varphi, z, \omega) = \sum_j \mathbf{u}_j(r, \varphi, z, \omega), \tag{5.16}$$

where

$$\mathbf{u}_j(r, \varphi, z, \omega) = \frac{1}{4\pi^2} \tag{5.17}$$

$$\times \int_0^{2\pi} \left(\int_{\Gamma^+(\gamma)} \mathbf{K}(\alpha, \gamma, z, \omega) \widetilde{\mathbf{Q}}_j(\alpha, \gamma) \, e^{-i\alpha \left(r\cos(\gamma - \varphi) + g_j(\gamma) \right)} \alpha \, d\alpha \right) d\gamma.$$

For the computation of the term $\mathbf{u}_j(r, \varphi, z, \omega)$ an algorithm of the evaluation of the contour integral along $\Gamma^+(\gamma)$ is considered. It is based on Cauchy's theorem, where γ as well as z, ω, r, φ are considered as parameters. First, the intervals with respect to γ are determined, in which the integrand in (5.17) decreases in the upper or lower half-plane respectively. These intervals[1] are found to be $\gamma_j^{\pm}(r, \varphi)$, where

$$\cos(\gamma - \varphi) + \frac{g_j(\gamma)}{r} < 0 \text{ for } \gamma \in \gamma_j^+(r, \varphi),$$

$$\cos(\gamma - \varphi) + \frac{g_j(\gamma)}{r} > 0 \text{ for } \gamma \in \gamma_j^-(r, \varphi), \tag{5.18}$$

$$(0, 2\pi) = \gamma_k^+(r, \varphi) \cup \gamma_k^-(r, \varphi).$$

The exponents $\exp\left(-i\alpha \left(r\cos(\gamma - \varphi) + g_j(\gamma)\right)\right)$ in (5.17), corresponding to these two subsets, are decaying in the upper half-plane for $\gamma \in \gamma_j^+(r, \varphi)$ and in the lower half-plane for $\gamma \in \gamma_j^-(r, \varphi)$. It allows to rewrite the integral (5.17) as a sum

$$\mathbf{u}_j(r, \varphi) = \mathbf{u}_j^+(r, \varphi, z, \omega) + \mathbf{u}_j^-(r, \varphi, z, \omega),$$

---

[1] As a rule, the sets $\gamma_j^{\pm}(r, \varphi)$ are the sets of separate non-crossing.

where

$$\mathbf{u}_j^{\pm}(r,\varphi,z,\omega) \;=\; \frac{1}{4\pi^2}$$

$$\times \int\limits_{\gamma_j^{\pm}(r,\varphi)} \left(\int\limits_{\Gamma^+(\gamma)} \mathbf{K}(\alpha,\gamma,z,\omega)\widetilde{\mathbf{Q}}_j(\alpha,\gamma)\, e^{-i\alpha\left(r\cos(\gamma-\varphi)+g_j(\gamma)\right)}\alpha\, d\alpha \right) d\gamma. \tag{5.19}$$

Note that for loads given in section 3.1.4, the intervals $\gamma_j^{\pm}(r,\varphi)$ can be written in an explicit form. For example, for a point excitation source (see Equation (3.17)) it follows that

$$\gamma_1^+(r,\varphi) = (\varphi + \pi/2, \varphi + 3\pi/2), \; \gamma_1^-(r,\varphi) = (\varphi - \pi/2, \varphi + \pi/2). \tag{5.20}$$

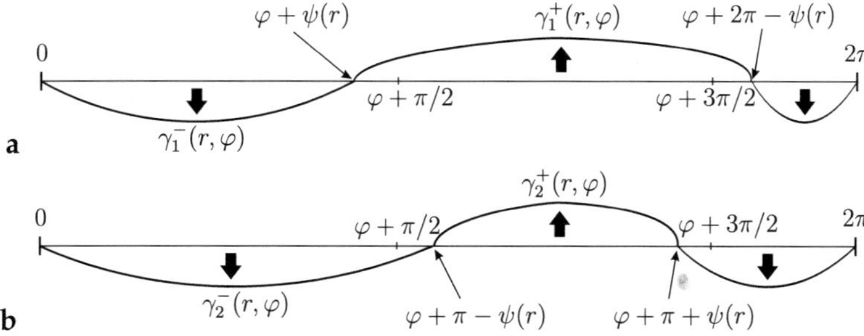

Figure 5.4: Decomposition of interval $(0, 2\pi)$ according to (5.18) for a circular piezo-electric wafer (3.17). (a) Intervals $\gamma_1^+(r,\varphi)$ and $\gamma_1^-(r,\varphi)$. (b) Intervals $\gamma_2^+(r,\varphi)$ and $\gamma_2^-(r,\varphi)$

For a circular piezo-actuator (3.17) the intervals

$$\begin{aligned}
\gamma_1^+(r,\varphi) &= (\varphi + \psi(r), \varphi + 2\pi - \psi(r)), \\
\gamma_1^-(r,\varphi) &= (\varphi + \pi - \psi(r), \varphi + \pi + \psi(r)), \\
\gamma_2^+(r,\varphi) &= (\varphi - \psi(r), \varphi + \psi(r)), \\
\gamma_2^-(r,\varphi) &= (\varphi - \pi + \psi(r), \varphi + \pi - \psi(r))
\end{aligned} \tag{5.21}$$

are obtained, where $\psi(r) = \arccos(A/r)$. These intervals are schematically depicted in Figure 5.4. For a rectangular piezo-actuator and MFC piezo-actuator described by pin-force models in a wavenumber domain by formulas (3.19) and (3.20) respectively, the intervals are

$$\begin{aligned}
\gamma_j^+(r,\varphi) &= (\theta_j(r,\varphi) + \pi/2, \theta_j(r,\varphi) + 3\pi/2), \\
\gamma_j^-(r,\varphi) &= (\theta_j(r,\varphi) - \pi/2, \theta_j(r,\varphi) + \pi/2)
\end{aligned} \tag{5.22}$$

similar to [109] are obtained, where $j = 1, 2, 3, 4$ and

$$
\begin{aligned}
\theta_1(r, \varphi) &= \arctan\left(\frac{r \sin \varphi - A_2}{r \cos \varphi - A_1}\right), \quad
\theta_2(r, \varphi) = \arctan\left(\frac{r \sin \varphi - A_2}{r \cos \varphi + A_1}\right), \\
\theta_3(r, \varphi) &= \arctan\left(\frac{r \sin \varphi + A_2}{r \cos \varphi - A_1}\right), \quad
\theta_4(r, \varphi) = \arctan\left(\frac{r \sin \varphi + A_2}{r \cos \varphi + A_1}\right).
\end{aligned}
\tag{5.23}
$$

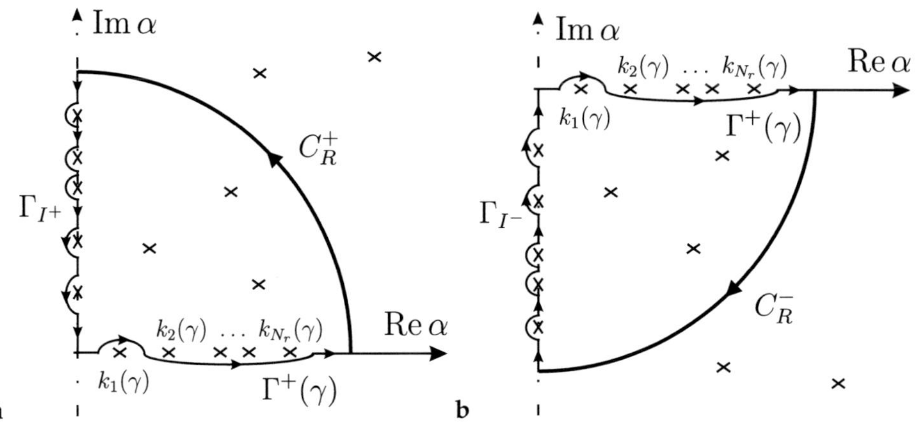

Figure 5.5: Contour $\Gamma^+(\gamma) \cup C_R^{\pm} \cup \Gamma_{I\pm}$ resulting after closing the contour $\Gamma^+(\gamma)$ upwards (a) or downwards (b)

As the sets $\gamma_j^{\pm}(r, \varphi)$ are determined, the contour $\Gamma^+(\gamma)$ is closed upwards into the first quadrant of α complex plane for $\gamma \in \gamma_k^+(r, \varphi)$ and downwards into the fourth quadrant of α complex plane for $\gamma \in \gamma_k^-(r, \varphi)$ by adding to the contour $\Gamma^+(\gamma)$ the contours $C_R^{\pm}$ in form of a corresponding quarter of a circle with radius R ($R \to \infty$) and $\Gamma_{I\pm}$, i.e. contours going along the positive and negative imaginary semi-axes, respectively. The resulting closed contour[1] is shown in Figure 5.5. According to Cauchy's residue theorem [3], the integral over the closed contour $\Gamma^+(\gamma) \cup C_R^{\pm} \cup \Gamma_{I\pm}$ equals to the sum of the residues evaluated in poles located inside of the contour times factor $2\pi \mathrm{i}$, i.e. the integral (5.19) can be evaluated as

$$
\begin{aligned}
\mathbf{u}_j^{\pm}(r, \varphi, z, \omega) &= \pm \frac{\mathrm{i}}{2\pi} \sum_{m=1}^{\infty} \int\limits_{\gamma_j^{\pm}(r,\varphi)} \mathbf{b}_{j,m}^{\pm}(\gamma, z, \omega)\, e^{-\mathrm{i}k_m^{\pm}(\gamma)(r \cos(\gamma - \varphi))}\, \mathrm{d}\gamma \\
&\quad - \mathbf{d}_j^{\pm}(r, \varphi, z, \omega),
\end{aligned}
\tag{5.24}
$$

---

[1]This contour corresponds to the case of the plate with free upper and lower boundaries, whereas for a half-space and for a plate with a clamped lower boundary, the branch cuts at branch points should be taken into account.

where it is taken into account that integrals over the quadrants $C_R^{\pm}$ are equal to zero (see proof in [109]). The functions $\mathbf{b}_{j,m}^{(\pm)}(\gamma, z, \omega)$ are expressed in terms of residues of Green's matrix $\mathbf{K}(\alpha, \gamma, z, \omega)$, namely

$$\mathbf{b}_{j,m}^{\pm}(\gamma, z, \omega) = \operatorname{res} \mathbf{K}(\alpha, \gamma, z, \omega)\Big|_{\alpha = k_m^{\pm}(\gamma)} \mathbf{Q}_j(k_m^{\pm}(\gamma), \gamma) k_m^{\pm}(\gamma). \tag{5.25}$$

The poles $k_m^{+}(\gamma)$ correspond to the real regular, complex and purely imaginary poles located above the contour $\Gamma^{+}(\gamma)$ in the first quadrant, whereas the poles $k_m^{-}(\gamma)$ are irregular real poles and the complex poles located below the contour $\Gamma^{+}(\gamma)$ in the fourth quadrant. In an implementation on PC, the residues can be evaluated with a high accuracy according to (B.1) given in Appendix B.1.

The functions $\mathbf{d}_j^{\pm}(r, \varphi, z, \omega)$ represent the integrals over contours $\Gamma_{I\pm}$ which are going in the complex plane α along the imaginary positive and negative semi-axes, i.e. $\operatorname{Re}\alpha = 0$, $\operatorname{Im}\alpha \geq 0$ and $\operatorname{Im}\alpha \leq 0$ for Γ_{I+} and Γ_{I-}, respectively. The contours $\Gamma_{I\pm}$ are deviating from the imaginary axis in the complex plane in order to bypass the purely imaginary poles of Green's matrix. Due to the condition on α: $\operatorname{Im}\alpha > 0$ if $\operatorname{Re}\alpha = 0$, the contours bypass all positive pure imaginary poles from the left (deviating into the second quadrant) and all negative pure imaginary poles from the right (deviating into the fourth quadrant), respectively, as it is shown in Figure 5.5. However, such contours are chosen for convenience since the radiation principles do not define any limitations on the choice of the integration paths over the imaginary axis. The corresponding integrals over these contours are defined as

$$\mathbf{d}_j^{\pm}(r, \varphi, z, \omega) = \frac{1}{4\pi^2} \int\limits_{\gamma_j^{\pm}(r,\varphi)} \left(\int\limits_{\Gamma_{I\pm}} \mathbf{K}(\alpha, \gamma, z, \omega) \mathbf{Q}_j(\alpha, \gamma) \, e^{-i\alpha r \cos(\gamma - \varphi)} \alpha \, d\alpha \right) d\gamma. \tag{5.26}$$

Note that Equation (5.24) requires the numerical integration with respect to γ for the terms corresponding to the contributions of residues and the evaluation of double integral (5.26), which is generally simpler because of the higher rate of decay as for an initial integral (5.1). In the following the evaluation of displacement applying the Equation (5.24) is called *Residue Integration Technique* (RIT).

Note that the formula (5.24) for the computation of the displacement vector $\mathbf{u}(r, \varphi, z, \omega)$ is similar to those obtained in [109] for partial cases of loads (3.17), (3.19). The main difference lies in the fact that in the approach applied in [109], the integrals along the imaginary semi-axes (5.26) are not taken into account. However, in spite of the exponential decay of the integrands at all values of γ (except of the endpoints of intervals $\gamma_j^{\pm}(r, \varphi)$) while integrating with respect to α, the contribution of these integrals can considerably influence the solution of the problem especially in a near-field to an excitation source. This is the reason why the formula (5.24) is not convenient for practical applications. However, the representation obtained in this section can essentially be simplified, as it is shown in the next sections.

5.3.2 Analysis of the solution of the problem in the far-field. Contribution of integrals along the imaginary semi-axes

In this section the formula (5.24) is considered if $r \to \infty$ is assumed. Due to the independence of $g_j(\gamma)$ on r (see Equation (5.18)), all intervals $\gamma_j^\pm(r, \varphi)$ are found to be independent of j, namely

$$
\begin{aligned}
\gamma_j^+(r,\varphi) \to \gamma^+(\varphi) &= \left(\varphi + \frac{\pi}{2}, \varphi + \frac{3\pi}{2}\right), \\
\gamma_j^-(r,\varphi) \to \gamma^-(\varphi) &= \left(\varphi - \frac{\pi}{2}, \varphi + \frac{\pi}{2}\right), \\
\text{i.e.} \quad \gamma^\pm(\varphi) &= \left(\varphi \pm \frac{\pi}{2}, \varphi + \pi \pm \frac{\pi}{2}\right).
\end{aligned}
\tag{5.27}
$$

This means that at sufficiently large distances, the influence of the type of the load on the intervals, where the contour is closed upwards or downwards, is neglible and it leads to the following representation for the displacement vector for $r \to \infty$:

$$
\mathbf{u}(r, \varphi, z, \omega) \sim \mathbf{u}^+(r, \varphi, z, \omega) + \mathbf{u}^-(r, \varphi, z, \omega),
\tag{5.28}
$$

$$
\mathbf{u}^\pm(r, \varphi, z, \omega) = \pm \frac{i}{2\pi} \sum_{m=1}^\infty \int_{\gamma^\pm(\varphi)} \mathbf{b}_m^\pm(\gamma, z, \omega)\, e^{-ik_m^\pm(\gamma) r \cos(\gamma - \varphi)}\, d\gamma - \mathbf{d}^\pm(r, \varphi, z, \omega),
$$

where

$$
\mathbf{b}_m^\pm(\gamma, z, \omega) = \operatorname{res} \mathbf{K}(\alpha, \gamma, z, \omega)\Big|_{\alpha = k_m^\pm(\gamma)} \mathbf{Q}(k_m^\pm(\gamma), \gamma) k_m^\pm(\gamma),
\tag{5.29}
$$

and

$$
\mathbf{d}^\pm = \left(d_1^\pm, d_2^\pm, d_3^\pm\right), \quad d_i^\pm = \sum_{j=1}^3 d_{ij}^\pm,
\tag{5.30}
$$

$$
d_{ij}^\pm(r, \varphi, z, \omega) = \frac{1}{4\pi^2} \int_{\Gamma_{I^\pm}} \left(\int_{\varphi \pm \pi/2}^{\varphi + \pi \pm \pi/2} K_{ij}(\alpha, \gamma, z, \omega) Q_j(\alpha, \gamma)\, e^{-i\alpha r \cos(\gamma - \varphi)}\, d\gamma \right) \alpha\, d\alpha.
$$

Note that the terms equal to (5.30) occur in the research works [40, 43, 69].

In terms $\mathbf{d}^\pm(r, \varphi, z, \omega)$ the method of stationary phase (see Appendix B.2) is applied to the integral[1] with respect to γ. This yields the representation for the integral along

---

[1]In the following the dependence of the integrands on parameters z and ω is omitted for simplicity.

the positive imaginary semi-axis

$$
\begin{aligned}
d_{ij}^{+}(r,\varphi) &= \frac{1}{4\pi^2} \int_{\Gamma_{I+}} \alpha \left(\frac{1}{i\alpha r} \left(-K_{ij}\left(\alpha,\varphi+\frac{3\pi}{2}\right) Q_j\left(\alpha,\varphi+\frac{3\pi}{2}\right) \right.\right. \\
&\quad - \left.\left. K_{ij}\left(\alpha,\varphi+\frac{\pi}{2}\right) Q_j\left(\alpha,\varphi+\frac{\pi}{2}\right) \right) d\alpha + \mathcal{O}\left(\alpha^{-1}r^{-2}\right) \right) \\
&= \frac{i}{4\pi^2 r} \int_{\Gamma_{I+}} \left(K_{ij}\left(\alpha,\varphi+\frac{3\pi}{2}\right) Q_j\left(\alpha,\varphi+\frac{3\pi}{2}\right) \right. \\
&\quad + \left. K_{ij}\left(\alpha,\varphi+\frac{\pi}{2}\right) Q_j\left(\alpha,\varphi+\frac{\pi}{2}\right) \right) d\alpha + \mathcal{O}\left(r^{-2}\right),
\end{aligned}
\tag{5.31}
$$

where $i,j = 1,2,3$. Then, the corresponding integral along the negative imaginary semi-axis is obtained for $i,j = 1,2,3$ as[1]

$$
\begin{aligned}
d_{ij}^{-}(r,\varphi) &= -\frac{i}{4\pi^2 r} \int_{\Gamma_{I-}} \left(K_{ij}\left(\alpha,\varphi+\frac{\pi}{2}\right) Q_j\left(\alpha,\varphi+\frac{\pi}{2}\right) \right. \\
&\quad + \left. K_{ij}\left(\alpha,\varphi+\frac{3\pi}{2}\right) Q_j\left(\alpha,\varphi+\frac{3\pi}{2}\right) \right) d\alpha + \mathcal{O}\left(r^{-2}\right).
\end{aligned}
\tag{5.32}
$$

Then, taking into account the fact that $\Gamma_{I-} \cup -\Gamma_{I+} = \Gamma_{I}$, where Γ_{I} is a contour mostly coinciding with the whole imaginary axis, the integrals in (5.31), (5.32) are combined for each $i,j = 1,2,3$ to one integral:

$$
\begin{aligned}
d_{ij}(r,\varphi) &= d_{ij}^{+}(r,\varphi) + d_{ij}^{-}(r,\varphi) = -\frac{i}{4\pi^2 r} \int_{\Gamma_{I}} \left(K_{ij}\left(\alpha,\varphi+\frac{\pi}{2}\right) \right. \\
&\quad \times\; Q_j\left(\alpha,\varphi+\frac{\pi}{2}\right) + K_{ij}\left(\alpha,\varphi+\frac{3\pi}{2}\right) Q_j\left(\alpha,\varphi+\frac{3\pi}{2}\right) \right) d\alpha + \mathcal{O}\left(r^{-2}\right).
\end{aligned}
\tag{5.33}
$$

The integrand decays at least as[2] α^{-1} in the whole complex plane as it follows from the properties of Green's matrix components (3.70). The contour Γ_{I} can be closed rightwards in the right half-plane. The application of the residue theorem brings the integrals (5.33) to the form

$$
\begin{aligned}
d_{ij}(r,\varphi) &= \frac{1}{2\pi r} \sum_{m=1}^{\infty} \left[\operatorname{res} K_{ij}\left(\alpha,\varphi+\frac{3\pi}{2}\right) Q_j\left(\alpha,\varphi+\frac{3\pi}{2}\right) \Bigg|_{\alpha=k_m\left(\varphi+\frac{3\pi}{2}\right)} \right. \\
&\quad + \left. \operatorname{res} K_{ij}\left(\alpha,\varphi+\frac{\pi}{2}\right) Q_j\left(\alpha,\varphi+\frac{\pi}{2}\right) \Bigg|_{\alpha=k_m\left(\varphi+\frac{\pi}{2}\right)} \right] + \mathcal{O}(r^{-2}),
\end{aligned}
\tag{5.34}
$$

---

[1] Using the periodicity of the functions $K_{ij}(\alpha,\gamma)$ and $Q_j(\alpha,\gamma)$ with respect to γ (period 2π).
[2] If the decay rate is exactly α^{-b}, $0 < b \leq 1$, then the result should be understood as a generalized function.

where the poles $k_m(\gamma)$ are located in the first and fourth quadrants (i.e. in the right half-plane) for both values of γ: $\varphi + \pi/2$ and $\varphi + 3\pi/2$.

Finally, for $r \to \infty$ the representation

$$\mathbf{u}(r, \varphi, z, \omega) \;\sim\; \sum_{m=1}^{\infty} \mathbf{u}_m^+ + \sum_{m=1}^{\infty} \mathbf{u}_m^- - \mathbf{d}(r, \varphi, z, \omega), \tag{5.35}$$

$$\mathbf{u}_m^\pm(r, \varphi, z, \omega) \;=\; \pm\frac{\mathrm{i}}{2\pi} \int\limits_{\varphi\pm\pi/2}^{\varphi+\pi\pm\pi/2} \mathrm{res}\,\mathbf{K}(\alpha, \gamma, z, \omega)\mathbf{Q}_j(\alpha, \gamma)\bigg|_{\alpha=k_m^\pm(\gamma)} \mathrm{e}^{-\mathrm{i}k_m^\pm(\gamma)r\cos(\gamma-\varphi)}\,\mathrm{d}\gamma$$

for the computation of displacements is derived, where the components of the vector function $\mathbf{d}(r, \varphi, z, \omega)$ are given by (5.34). The application of this modified formula (5.35) in the following is called *Far-Field Residue Integration Technique* (FFRIT).

Note that the displacement values $\mathbf{u}_m^\pm(r, \varphi, z, \omega)$ in (5.35) corresponding to an m-th pole in the I ("+") or IV ("−") quadrants should be not confused with the values $\mathbf{u}_j^\pm(r, \varphi, z, \omega)$ from (5.17) corresponding to the contribution of the j-th function $\mathbf{q}_j(\mathbf{x})$ in the representation (5.12).

Remark 5.5 *In case of symmetry of $K_{ij}(\alpha, \gamma)Q_j(\alpha, \gamma)$ with respect to* $\mathrm{Re}\,\alpha = 0$, *the term* $\mathcal{O}(r^{-2})$ *in (5.35) can be replaced by* $\mathcal{O}(r^{-3})$.

Again, as in case of the expression (5.24), here the evaluation of a single oscillating integral over γ is needed. However, expression (5.35) requires the integration over simple intervals $\gamma^\pm(\varphi)$ instead of more complex intervals $\gamma_j^\pm(r, \varphi)$ in (5.24). Moreover, the double integral over γ and $\Gamma_{I\pm}$ is replaced by its asymptotics and can be computed in an explicit form as (5.34).

5.3.3 Approximation of surface load as a sum of point sources

Below the formulas derived in section 5.3.2 are applied to the case of the load function given as a sum of point sources. Note that if the point source at the origin is considered, its $\gamma_j^\pm(r, \varphi)$ and $\gamma^\pm(\varphi)$ are equal, i.e. the far-field assumption (5.27) is exactly satisfied for all r. Hence, in a local polar coordinate system $(\tilde{r}_j, \tilde{\varphi}_j)$ associated with a point source at[1] (x_j, y_j) (see Figure 5.6) given as

$$\tilde{r}_j = \sqrt{(x - x_j)^2 + (y - y_j)^2}, \quad \cos\tilde{\varphi}_j = (x - x_j)/\tilde{r}_j, \quad \sin\tilde{\varphi}_j = (y - y_j)/\tilde{r}_j, \tag{5.36}$$

the displacement vector can be calculated at each point $(x, y) = (r\cos\varphi, r\sin\varphi)$, $(x, y) \neq (x_j, y_j)$ using the representation (5.35). In global coordinates the displacement

---

[1] Here all points have the same z-coordinate.

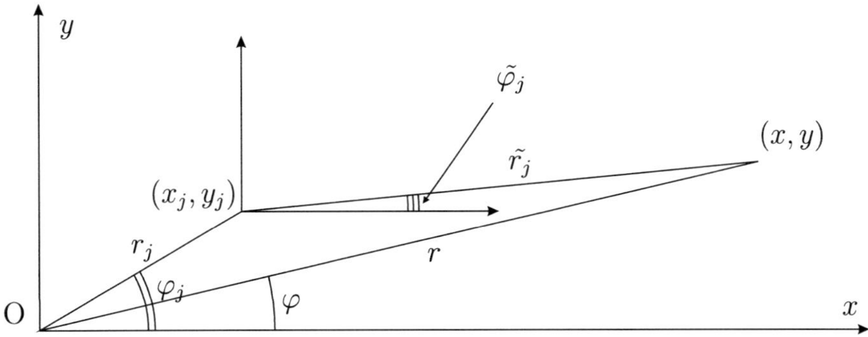

Figure 5.6: Local coordinate system associated with (x_j, y_j)

corresponding to the sum of point sources in form of (5.13) is found to be

$$\mathbf{u}(r, \varphi, z, \omega) \;=\; \delta_s^2 \sum_{j=1}^{N_q} \left[\sum_{m=1}^{\infty} \mathbf{u}_{j,m}^+ + \sum_{m=1}^{\infty} \mathbf{u}_{j,m}^- - \mathbf{d}_j(r, \varphi, z, \omega) \right] + \mathcal{O}(r^{-2}), \quad r \to \infty,$$

$$\mathbf{u}_{j,m}^{\pm}(r, \varphi, z, \omega) \;=\; \pm \frac{\mathrm{i}}{2\pi} \int_{\tilde{\varphi}_j \pm \pi/2}^{\tilde{\varphi}_j + \pi \pm \pi/2} \mathbf{b}_{j,m}^{\pm}(\gamma, z, \omega)\, \mathrm{e}^{-\mathrm{i}k_m^{\pm}(\gamma)\tilde{r}_j \cos(\gamma - \tilde{\varphi}_j)}\, \mathrm{d}\gamma,$$

$$\mathbf{b}_{j,m}^{\pm}(\gamma, z, \omega) \;=\; \operatorname{res} \mathbf{K}(\alpha, \gamma, z, \omega)\Big|_{\alpha = k_m^{\pm}(\gamma)} \mathbf{q}_j k_m^{\pm}(\gamma), \tag{5.37}$$

$$d_j(r, \varphi) \;=\; d_j^+ + d_j^-$$

$$= \; \frac{1}{2\pi r_j} \sum_{m=1}^{\infty} \left[\operatorname{res} \mathbf{K}\left(\alpha, \tilde{\varphi}_j + \frac{3\pi}{2}\right) \mathbf{q}_j \Big|_{\alpha = k_m\left(\tilde{\varphi}_j + \frac{3\pi}{2}\right)} \right.$$

$$\left. + \; \operatorname{res} \mathbf{K}\left(\alpha, \tilde{\varphi}_j + \frac{\pi}{2}\right) \mathbf{q}_j \Big|_{\alpha = k_m\left(\tilde{\varphi}_j + \frac{\pi}{2}\right)} \right] + \mathcal{O}(\tilde{r}_j^{-2}).$$

If the pointwise approximation (5.13) describes the excitation source with only minor errors, the former (5.37) gives the nearly exact solution of the corresponding wave propagation problem. This representation in the following is called *Pointwise Far-Field Residue Integration Technique* (P-FFRIT).

Remark 5.6 *At large distance r, in (5.36) the angle $\tilde{\varphi}_j \approx \varphi$ and $\tilde{r}_j \approx r$, i.e. the solution given by (5.37) in a far-field is turning to the far-field solution according to (5.35).*

5.3.4 Comparison of methods of displacement evaluation based on the residue theorem

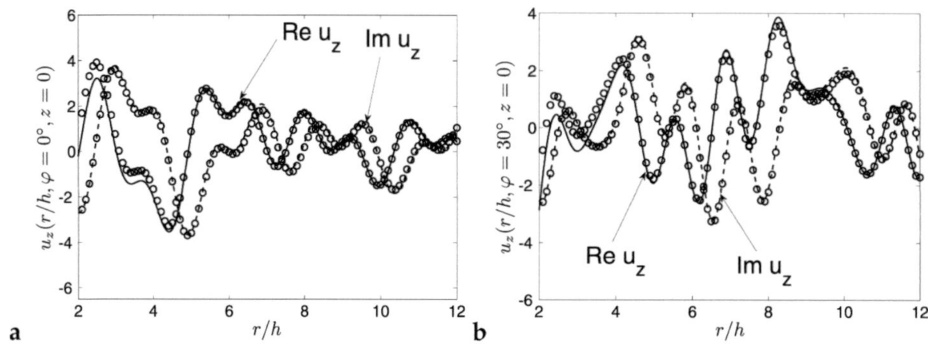

Figure 5.7: Comparison of complex displacements $u_z(r)$ obtained using the DCI (5.7) applying adaptive quadratures ($R = 50$, "$-$") and FFRIT (5.24), ("o") at $z = 0$, $\varphi = 0°$ (a) and $\varphi = 30°$ excited in a unidirectional plate made of graphite-epoxy I (Table A.1) at a frequency-thickness product $1.472\text{MHz} \cdot \text{mm}$ by a concentrated vertical point source (3.17) at $r = 0$

All the expressions obtained in previous sections for the displacement vector are based on the residue theorem and represent it as a sum of contributions of residues in the poles of Green's matrix, or, in other words, as a sum of Lamb wave modes. In (5.24) additionally the integrals over the imaginary semi-axes contribute to the displacements. As it is shown in section 5.3.2, these integrals contribute in the far-field as $\mathcal{O}(r^{-1})$ and they can be represented again a sum of contributions of residues in poles. It is concluded that in the formula (5.24) the integrals over $\Gamma_{I\pm}$ are ignored, as it is done by authors in [107, 109], the sum of contributions of residues even in a large number of poles, does not lead to more accurate result.

The modified far-field residue representation (5.35) takes into account the asymptotic expansion for the integrals over the imaginary semi-axes and represents the displacement solely as a sum of contributions of residues in poles of Green's matrix. It is proven to be more accurate than the former (5.24), at least in the far-field. However, the numerical error can grow if the complex poles with a large imaginary part $k_I > 0$ are considered, e.g. for the circular piezoelectric wafer, see Equation (3.17), the corresponding multiplier $J_1(A_o k_m^+) \sim \exp(A_o k_I)$ (considering the decomposition provided in the given formula). For values of $\gamma \in (\varphi + \pi/2, \varphi + \pi - \psi(r))$ according to (5.21), in the representation (5.35) the integral over γ needs to be evaluated numerically, whereas the integrand is taking large values which together with the numerical errors occuring in a numerical computation of Bessel function of the complex argument with a large imaginary part can lead to a significant error in the result, i.e. the use of a larger number of complex poles do not lead to a more accurate result even in the far-field. Moreover, this representation gives just the far-field approximation of the displacement problem and its use in the near-field may lead to wrong results.

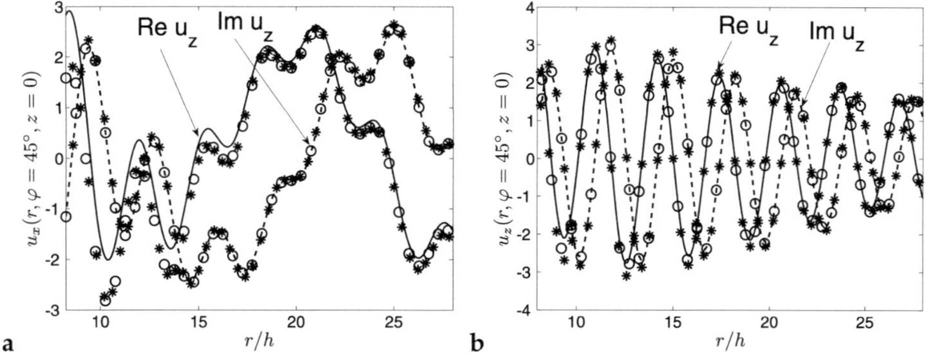

Figure 5.8: Comparison of $u_x(r)$ (a) and $u_z(r)$ (b) computed applying different approaches for $\varphi = 45°$ in case of the excitation of Lamb waves by MFC piezo-electric actuator (3.20), ($A_1/h = 8$, $A_2/h = 2$) at a frequency-thickness 500 KHz $\cdot$ mm in a cross-ply symmetric $[45_6/-45_6]_s$ plate made of AS4/3502 (Table A.1). Results of DCI (5.7) with $R = 25$ ("o"), of FFRIT ("−", Equation (5.35)) and of P-FFRIT with modelling of MFC by 21 point sources on each side ("∗", Equation (5.37)) are plotted

In contrast, the representation (5.37) allows to obtain accurate results in both the near- and the far-field by taking into account without any limitations so many complex poles as are known. Moreover, this approach is working well even for surface loads, for which the analytical wavenumber representation $\mathbf{Q}(\alpha, \gamma)$ is not available. On the other hand, this approach has several limitations, namely only the approximation of the load function is used, so that the accuracy of the modelling depends on the number of point sources considered. Moreover, it causes an additional computational time due to the calculation of $2 \cdot N_q$ integrals over γ against two integrals over γ in (5.35).

These properties of the approaches given in previous sections are discussed on several examples of numerical computations. A first example is given by the unidirectional plate of graphite-epoxy I under the excitation by a vertical point source, see Equation (3.17), at $r = 0$ at a frequency-thickness 1.472MHz $\cdot$ mm. The values of out-of-plane displacement component (real and imaginary parts) are calculated in directions $\varphi = 0°$ (Figure 5.7a) and (Figure 5.7b) for $z = 0$ applying the DCI according to (5.7) using adaptive quadratures ($R = 50$, "−") and FFRIT (see Equation (5.24), "o"), whereas in the residue integration only real poles are taken into account. Note that at this frequency in spite of regular poles also one irregular pole exists (see Figure 4.2). The results of both methods well coincide with each other and the difference between them decreases as r increases. However, in spite of good coincidence, the results of both methods are not exactly equal in near-field to the source.

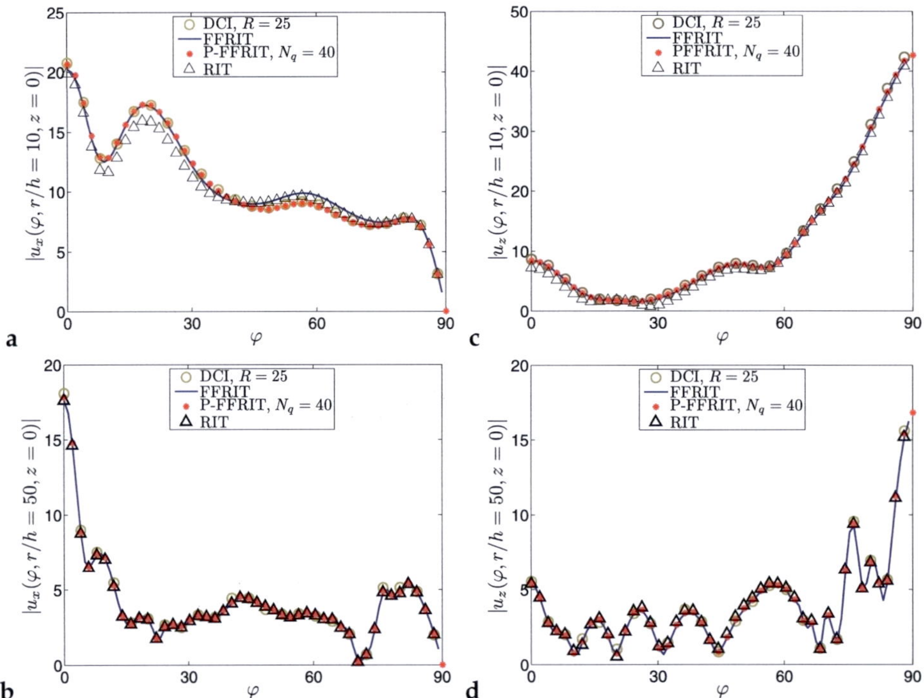

Figure 5.9: Comparison of $|u_x(\varphi)|$ (a and b) and $|u_z(\varphi)|$ (c and d) computed for $r/h = 10$ (a and c) and $r/h = 50$ (b and d) in case of the excitation of Lamb waves by a circular piezo-electric actuator of radius $A_o/h = 4$ (3.17) at a frequency-thickness 1000 KHz · mm in a cross-ply non-symmetric $[0/90/0/90]$ plate made of CFRP-T700GC/M21 (Table A.1)

Another example is given by the $[45_6/-45_6]_s$ cross-ply plate made of AS4/3502 (Table A.1), which is harmonically excited by an MFC piezo-electric actuator (see Equation (3.20)) with $A_1/h = 8$ and $A_2/h = 2$ at frequency-thickness 500 KHz · mm. The real and imaginary amplitudes of $u_x(r)$ (Figure 5.8a) and $u_z(r)$ (Figure 5.8b) are evaluated in direction $\varphi = 45°$ using three approaches: direct integration (DCI: see Equation (5.7)) for $R = 25$ - marked as "o", far-field residue integration technique (FFRIT: see Equation (5.35)) - plotted as straight lines, and the results of modelling of MFC by 21 point sources on each side (P-FFRIT: see Equation (5.37)) - marked as "*", respectively. In FFRIT and P-FFRIT, only contributions of all three real regular poles are taken into account. Due to the large near-field produced by a large transducer, the far-field approximation ("−") does not allow to reach a sufficient accuracy at distances below $r/h = 20$. The results of modelling by point sources, however, agree well with the results of direct integration at distances $r/h > 9$. At the same time, this technique requires much more computational time as the use of (5.35) since the sum with respect to j in Equation (5.37) should be computed for each value of r and φ, i.e. the computational time increases N_q times.

A similar comparison is performed for a cross-ply non-symmetric $[0/90/0/90]$ plate made of CFRP-T700GC/M21 (Table A.1) under harmonic excitation by a circular piezo-electric actuator of radius $A_o/h = 4$ (3.17) at a frequency-thickness 1000 KHz · mm (Figure 5.9). However, additionally the displacement vector is computed applying (5.24), where, as before, only the contributions of all five real regular poles are considered. Whereas at $r/h = 50$ (Figure 5.9b and d) no differences between the results of all approaches are visible, at $r/h = 10$ the approach without considering the integrals over imaginary semi-axes (5.24) shows for most values of φ the largest deviation of results from the most accurate approach, namely the DCI (5.7). The technique of modelling by point sources (5.37) coincides well with the direct integration, the results of far-field residue integration (5.35) are almost coinciding with the direct integration approach, however, at $r/h = 10$ this agreement is not perfect.

5.4 Analysis of expressions obtained for the example of an isotropic laminate

The far-field representation (5.35) of the displacement vector can be validated for the isotropic laminate, for which in case of the axis-symmetric excitation source an exact analytic solution is known [39, 108]. For example, the isotropic plate under an excitation by a circular piezo-actuator (3.17) of radius A_o, the following exact representation of the displacement field

$$u_r(r,z,\omega) = \frac{\tau_0}{2} \sum_{m=1}^{\infty} \operatorname{res} M(\alpha,z,\omega)\Big|_{\alpha=k_m} k_m J_1(k_m A_o) H_1^{(1)}(k_m r), \qquad (5.38)$$

$$u_z(r,z,\omega) = -\frac{\tau_0}{2} \sum_{m=1}^{\infty} \operatorname{res} S(\alpha,z,\omega)\Big|_{\alpha=k_m} J_1(k_m A_o) H_0^{(1)}(k_m r)$$

is obtained for $r > A_o$, where $M(\alpha,z,\omega)$ and $S(\alpha,z,\omega)$ are given in Appendix A.8, $H_n^{(1)}(\alpha)$ are the Hankel functions of the first kind, and where the poles k_m lie in the upper half-plane.

In the following the solution is analyzed in the form (5.35) for the isotropic case. It is obtained in the wavenumber domain as

$$\mathcal{F}_{xy}[\mathbf{u}(x,y,z,\omega)] = \mathbf{K}(\alpha,\gamma,z,\omega)\mathbf{Q}(\alpha,\gamma) = i\tau_0 J_1(\alpha A_o) \begin{pmatrix} -i\cos\gamma\, M(\alpha,z,\omega) \\ -i\sin\gamma\, M(\alpha,z,\omega) \\ S(\alpha,z,\omega)/\alpha \end{pmatrix}. \quad (5.39)$$

The radial in-plane displacement component in Fourier domain is found to be

$$\begin{aligned}
\mathcal{F}_{xy}[u_r(x,y,z,\omega)] &= \mathcal{F}_{xy}[u_x(x,y,z,\omega)]\cos\varphi + \mathcal{F}_{xy}[u_y(x,y,z,\omega)]\sin\varphi \quad (5.40)\\
&= \tau_0 J_1(\alpha A_o) M(\alpha,z,\omega)\cos(\gamma-\varphi),
\end{aligned}$$

which by substituting it in (5.35) leads to the expressions

$$\begin{aligned}
u_r(r,z,\omega) &= \frac{i\tau_0}{2\pi}\left(\sum_{m=1}^{\infty} J_1(k_m^+ A_o)\left(\operatorname{res} M(\alpha,z,\omega)\Big|_{\alpha=k_m^+}\, k_m^+ \right) \int_{\varphi+\pi/2}^{\varphi+3\pi/2}\cos(\gamma-\varphi) \right.\\
&\quad \times\ e^{-ik_m^+ r\cos(\gamma-\varphi)}\,\mathrm{d}\gamma - \sum_{m=1}^{\infty} J_1(k_m^- A_o)\left(\operatorname{res} M(\alpha,z,\omega)\Big|_{\alpha=k_m^-}\, k_m^- \right)\\
&\quad \left. \times \int_{\varphi-\pi/2}^{\varphi+\pi/2}\cos(\gamma-\varphi)\, e^{-ik_m^- r\cos(\gamma-\varphi)}\,\mathrm{d}\gamma \right) + d_r(r,z,\omega), \quad (5.41)
\end{aligned}$$

$$\begin{aligned}
u_z(r,z,\omega) &= -\frac{\tau_0}{2\pi}\left(\sum_{m=1}^{\infty} J_1(k_m^+ A_o)\operatorname{res} S(\alpha,z,\omega)\Big|_{\alpha=k_m^+}\int_{\varphi+\pi/2}^{\varphi+3\pi/2}\cos(\gamma-\varphi) \right.\\
&\quad \times\ e^{-ik_m^+ r\cos(\gamma-\varphi)}\,\mathrm{d}\gamma\\
&\quad - \sum_{m=1}^{\infty} J_1(k_m^- A_o)\operatorname{res} S(\alpha,z,\omega)\Big|_{\alpha=k_m^+}\\
&\quad \left. \times \int_{\varphi-\pi/2}^{\varphi+\pi/2}\cos(\gamma-\varphi)\, e^{-ik_m^- r\cos(\gamma-\varphi)}\,\mathrm{d}\gamma \right) + d_z(r,z,\omega),
\end{aligned}$$

where the poles k_m are located in the right half-plane and according to (5.34) following relation holds

$$\begin{aligned}
\begin{pmatrix} d_r(r,z,\omega) \\ d_z(r,z,\omega) \end{pmatrix} &= \frac{\tau_0}{2\pi r}\sum_{m=1}^{\infty}\left(J_1(k_m A_o)\operatorname{res}\begin{pmatrix} M(\alpha,z,\omega)\cos(3\pi/2) \\ S(\alpha,z,\omega)/\alpha \end{pmatrix}\right)\Bigg|_{\alpha=k_m}\\
&\quad + \operatorname{res}\begin{pmatrix} M(\alpha,z,\omega)\cos(\pi/2) \\ S(\alpha,z,\omega)/\alpha \end{pmatrix}\Bigg|_{\alpha=k_m}\Bigg) + \mathcal{O}(r^{-2}) \quad (5.42)\\
&= \frac{i\tau_0}{\pi r}\sum_{m=1}^{\infty} J_1(k_m A_o)\operatorname{res}\begin{pmatrix} 0 \\ S(\alpha,z,\omega)/\alpha \end{pmatrix}\Bigg|_{\alpha=k_m} + \mathcal{O}(r^{-2}).
\end{aligned}$$

The integrals with respect to γ in (5.41) can be evaluated analytically applying formulas (B.18), (B.20) from Appendix B.3. This yields

$$
\begin{aligned}
u_r(r,z,\omega) &= \frac{\tau_0}{2}\left(\sum_{m=1}^{\infty} J_1(k_m^+ A_o)\,\mathrm{res}\,M(\alpha,z,\omega)\Big|_{\alpha=k_m^+}\; k_m^+ \cdot \big(J_1(k_m^+ r)\right.\\
&\quad + \mathrm{i}\left(\mathbf{H}_1(k_m^+ r) - \frac{2}{\pi}\right)\Big) - \sum_{m=1}^{\infty} J_1(k_m^- A_o)\,\mathrm{res}\,M(\alpha,z,\omega)\Big|_{\alpha=k_m^-}\; k_m^- \cdot \big(J_1(k_m^- r)\\
&\quad \left. - \mathrm{i}\left(\mathbf{H}_1(k_m^- r) - \frac{2}{\pi}\right)\right)\right) + d_r(r,z,\omega),
\end{aligned}
\tag{5.43}
$$

$$
\begin{aligned}
u_z(r,z,\omega) &= \frac{-\tau_0}{2}\left(\sum_{m=1}^{\infty} J_1(k_m^+ A_o)\,\mathrm{res}\,S(\alpha,z,\omega)\Big|_{\alpha=k_m^+}\; \big(J_0(k_m^+ r)+\mathrm{i}\mathbf{H}_0(k_m^+ r)\big)\right.\\
&\quad \left. - \sum_{m=1}^{\infty} J_1(k_m^- A_o)\,\mathrm{res}\,S(\alpha,z,\omega)\Big|_{\alpha=k_m^+}\; \big(J_0(k_m^- r)-\mathrm{i}\mathbf{H}_0(k_m^- r)\big)\right) + d_z(r,z,\omega),
\end{aligned}
$$

where $J_n(\alpha r)$ are Bessel functions and $\mathbf{H}_n(\alpha r)$ denote Struve functions[1] (see Equations (B.19), (B.21)). Taking into account the asymptotic properties of Bessel and Struve functions (B.14), (B.23) leads to the asymptotic expression

$$
\begin{aligned}
u_r(r,z,\omega) &= \frac{\tau_0}{2}\left(\sum_{m=1}^{\infty} J_1(k_m^+ A_o)\,\mathrm{res}\,M(\alpha,z,\omega)\Big|_{\alpha=k_m^+}\; k_m^+ \cdot \left(H_1^{(1)}(k_m^+ r) + \frac{2\mathrm{i}}{\pi(k_m^+)^2 r^2}\right)\right.\\
&\quad \left. - \sum_{m=1}^{\infty} J_1(k_m^- A_o)\,\mathrm{res}\,M(\alpha,z,\omega)\Big|_{\alpha=k_m^-}\; k_m^- \left(H_1^{(2)}(k_m^- r) - \frac{2\mathrm{i}}{\pi(k_m^-)^2 r^2}\right)\right)\\
&\quad + d_r(r,z,\omega) + \mathcal{O}(r^{-2}),
\end{aligned}
\tag{5.44}
$$

$$
\begin{aligned}
u_z(r,z,\omega) &= \frac{-\tau_0}{2}\left(\sum_{m=1}^{\infty} J_1(k_m^+ A_o)\,\mathrm{res}\,S(\alpha,z,\omega)\Big|_{\alpha=k_m^+}\; \left(H_0^{(1)}(k_m^+ r) + \frac{2\mathrm{i}}{\pi(k_m^+)r}\right)\right.\\
&\quad \left. - \sum_{m=1}^{\infty} J_1(k_m^- A_o)\,\mathrm{res}\,S(\alpha,z,\omega)\Big|_{\alpha=k_m^+}\; \left(H_0^{(2)}(k_m^- r) - \frac{2\mathrm{i}}{\pi k_m^- r}\right)\right)\\
&\quad + d_z(r,z,\omega) + \mathcal{O}(r^{-3}),
\end{aligned}
$$

for (5.43), where $H_n^{(j)}(\alpha r)$, $j = 1,2$ are the Hankel functions of the first and second kind, respectively. Taking into account asymptotic expressions for the integrals along

---

[1] Struve functions are scalar functions. However, as an exception they are traditionally denoted using bold font.

the imaginary axes (5.42) in (5.44) leads to expressions for the displacements:

$$
\begin{aligned}
u_r(r,z,\omega) &= \frac{\tau_0}{2}\left(\sum_{m=1}^{\infty} J_1(k_m^+ A_o)\,\mathrm{res}\,M(\alpha,z,\omega)\Big|_{\alpha=k_m^+} k_m^+\, H_1^{(1)}(k_m^+ r)\right. \\
&\quad \left. - \sum_{m=1}^{\infty} J_1(k_m^- A_o)\,\mathrm{res}\,M(\alpha,z,\omega)\Big|_{\alpha=k_m^-} k_m^-\, H_1^{(2)}(k_m^- r)\right) + \mathcal{O}(r^{-2}), \quad (5.45) \\
u_z(r,z,\omega) &= \frac{-\tau_0}{2}\left(\sum_{m=1}^{\infty} J_1(k_m^+ A_o)\,\mathrm{res}\,S(\alpha,z,\omega)\Big|_{\alpha=k_m^+} H_0^{(1)}(k_m^+ r)\right. \\
&\quad \left. - \sum_{m=1}^{\infty} J_1(k_m^- A_o)\,\mathrm{res}\,S(\alpha,z,\omega)\Big|_{\alpha=k_m^+} H_0^{(2)}(k_m^- r)\right) + \mathcal{O}(r^{-2}).
\end{aligned}
$$

The only difference which remains between the far-field representation (5.45) and the exact solution (5.38) is that the poles in (5.45) lie in the first and fourth quadrants instead of the first and second quadrants in (5.38). However, due to the symmetry of the poles $k_{m,\mathrm{IV}} = -k_{m,\mathrm{II}}$ (see section 4.1.1) and taking into account the properties of Bessel (B.13) and Hankel (B.16) functions as well as the properties of residues for even functions $M(\alpha,z,\omega)$ and $S(\alpha,z,\omega)$ (B.2), the far-field representations (5.35) of displacements in the isotropic laminate for an axis-symmetric excitation at the surface are finally written for $r \to \infty$ as

$$
u_r(r,z,\omega) = \frac{\tau_0}{2}\sum_{m=1}^{\infty} J_1(k_m A_o)k_m\,\mathrm{res}\,M(\alpha,z,\omega)\Big|_{\alpha=k_m} H_1^{(1)}(k_m r) + \mathcal{O}(r^{-2}), \qquad (5.46)
$$

$$
u_z(r,z,\omega) = -\frac{\tau_0}{2}\sum_{m=1}^{\infty} J_1(k_m A_o)\,\mathrm{res}\,S(\alpha,z,\omega)\Big|_{\alpha=k_m} H_0^{(1)}(k_m r) + \mathcal{O}(r^{-2}),
$$

where $\mathrm{Im}\,k_m \geq 0$ (k_m lies in an upper half-plane above the contour Γ). It is concluded that for $r \to \infty$ the difference between the exact analytical representation (5.38) and the representation (5.35) derived in section 5.3.2 is, as it was estimated, of order $\mathcal{O}(r^{-2})$.

Remark 5.7 *In fact, it can be proven that the expression for $u_z(r,z,\omega)$ is of order $\mathcal{O}(r^{-3})$.*

In Figure 5.10a the magnitude of the out-of-plane displacement is plotted in dependence on r, where the isotropic steel plate is excited by a circular piezoelectric wafer of radius $A_o/h = 4$ (3.17) at 850 KHz · mm. Here, the displacement is computed using the analytical formula (5.38) taking into account the contribution of all real poles and two nearest to the real-axis complex poles ($\mathrm{Im}\,k_c \approx 0.6$). In Figure 5.10a the good coincidence of both results is observed for $r/h > 13$. Note that according to formula (5.2) the near-field is defined as $r/h \leq 4^2 \cdot 5/(2\pi) \approx 13$. In Figure 5.10b only the contribution of these two complex poles is presented, however, additionally to the analytical representation (5.38), ("$-$") their contribution is calculated using the corresponding terms in formula (5.35) ("o") and in formula (5.24) ("$--$"). The results of applying the far-field representation (5.35) coincide with analytical results already at $r/h > 5$,

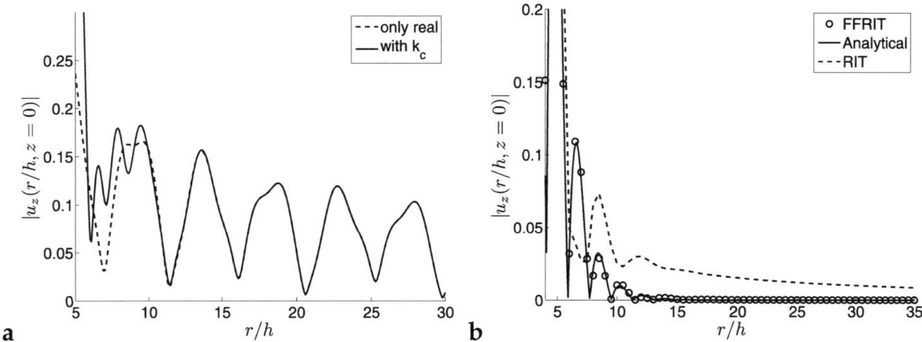

Figure 5.10: (a) Magnitudes $|u_z(r)|$ occuring in a steel plate at 850 KHz $\cdot$ mm under excitation by a circular piezoelectric wafer of radius $A_o/h = 4$, see Equation (3.17), computed using the exact analytical formula (5.38) taking into account all real poles and two nearest to the real axis complex poles (k_c and $-k_c^*$). (b) Contribution of complex (k_c and $-k_c^*$) poles into $|u_z(r)|$ computed applying different approaches

whereas the results corresponding to the more general representation (5.24) do not agree with an asymptotic solution. One can explain it by ignoring[1] the contribution of integrals over the imaginary semi-axes in (5.24) against considering it as a far-field estimation in (5.35).

5.5 Far-field asymptotics of the displacement vector

5.5.1 General asymptotic expansion

The expressions obtained for the displacement vector in section 5.3 give the acceptable accuracy often already in the near-field to the excitation source and allow to obtain an exact result in the far-field. However, the time needed for the evaluation in the far-field increases due to the oscillation of the integrands with respect to γ for growing r. Nevertheless, in practical applications of SHM mainly the displacements far away from the excitation source, where $r \gg A_o^2 k / (2\pi)$ are needed (A_0 describes the minimal radius of circle, which contains the bounded domain Ω, where the surface load is applied; k is a maximal wavenumber of propagating wave modes for the frequency considered). For the evaluation of integrals over γ in (5.35) for $r \to \infty$, the stationary phase method is applied (see Appendix B.2) similarly as it is performed in section 5.3.2 for an asymptotic analysis of expressions corresponding to the integration over γ and

---

[1]In the far-field according to section 5.3.2, the integrals over imaginary semi-axes can be represented as a sum of contribution of poles.

over the imaginary semi-axes. Then, the following integrals are considered

$$\mathbf{u}_m^\pm(r,\varphi,z,\omega) = \pm \frac{\mathrm{i}}{2\pi} \int\limits_{\varphi\pm\pi/2}^{\varphi+\pi\pm\pi/2} \mathbf{b}_m^\pm(\gamma,z,\omega)\, \mathrm{e}^{-\mathrm{i}k_m^\pm(\gamma)r\cos(\gamma-\varphi)}\, \mathrm{d}\gamma \tag{5.47}$$

for $r \to \infty$. The contribution of the pole $k_m^\pm(\gamma)$ into the integral (5.47) is brought by the endpoints of the interval $(\varphi \pm \pi/2,\, \varphi + \pi \pm \pi/2)$ and by the internal stationary points of the *phase function*

$$P_m^\pm(\gamma,\varphi) = -k_m^\pm(\gamma)\cos(\gamma - \varphi). \tag{5.48}$$

First, the contribution $\mathbf{e}_m^\pm(r,\varphi,z,\omega)$ of the endpoints of $\gamma^\pm(\varphi)$ is found to be[1]

$$
\begin{aligned}
\mathbf{e}_m^\pm(r,\varphi,z,\omega) &= \pm\frac{\mathrm{i}}{2\pi}\frac{1}{r}\left(\frac{\mathbf{b}_m^\pm\left(\varphi+\pi\pm\dfrac{\pi}{2},z,\omega\right)}{\mathrm{i}P_m'\left(\varphi+\pi\pm\dfrac{\pi}{2}\right)} - \frac{\mathbf{b}_m^\pm\left(\varphi\pm\dfrac{\pi}{2},z,\omega\right)}{\mathrm{i}P_m'\left(\varphi\pm\dfrac{\pi}{2}\right)} \right) \\
&\quad + \ \mathcal{O}(r^{-2}) \\[2mm]
&= -\frac{1}{2\pi r k_m^\pm\left(\varphi+\dfrac{\pi}{2}\right)}\left(\mathbf{b}_m^\pm\left(\varphi+\dfrac{3\pi}{2},z,\omega\right) + \mathbf{b}_m^\pm\left(\varphi+\dfrac{\pi}{2},z,\omega\right) \right) \\
&\quad + \ \mathcal{O}(r^{-2}).
\end{aligned}
\tag{5.49}
$$

Then, according to (5.35) the sum of all of contributions of endpoints of all poles in the right half-plane is brought to the form

$$
\begin{aligned}
\sum_{m_1=1}^{\infty} \mathbf{e}_{m_1}^+ + \sum_{m_2=1}^{\infty} \mathbf{e}_{m_2}^- &= -\frac{1}{2\pi r} \\
&\times \sum_{m=1}^{\infty} \left(\operatorname{res} \mathbf{K}\left(\alpha,\varphi+\frac{3\pi}{2},z,\omega\right)\Bigg|_{\alpha=k_m\left(\varphi+\frac{\pi}{2}\right)} \mathbf{Q}\left(k_m\left(\varphi+\frac{\pi}{2}\right),\varphi+\frac{3\pi}{2}\right) \right. \\
&\qquad\quad \left. + \operatorname{res} \mathbf{K}\left(\alpha,\varphi+\frac{\pi}{2},z,\omega\right)\Bigg|_{\alpha=k_m\left(\varphi+\frac{\pi}{2}\right)} \mathbf{Q}\left(k_m\left(\varphi+\frac{\pi}{2}\right),\varphi+\frac{\pi}{2}\right) \right) + \mathcal{O}(r^{-2}).
\end{aligned}
\tag{5.50}
$$

This expression is the same as (5.34), but with an opposite sign. One can see from the asymptotic expansion of (5.35) for $r \to \infty$, these terms compensate each other and only the contributions of the internal stationary points of phase function (5.48) remain.

For a real-valued or pure imaginary-valued pole $k_m^\pm(\gamma)$, its contribution is given by the stationary points $\gamma_{mp}^\pm(\varphi)$ satisfying the so called *phase equation*

$$\frac{k_{m,\gamma}'^\pm(\gamma)}{k_m^\pm(\gamma)} = \tan(\gamma - \varphi), \quad \gamma \in (\varphi \pm \pi/2,\, \varphi + \pi \pm \pi/2). \tag{5.51}$$

---

[1] Using the periodicity of the function $\mathbf{b}_m^\pm(\gamma,z,\omega)$ with respect to γ (period 2π).

Note that this equation can be rewritten using the definition of the steering angle (2.57) as

$$\varphi_m - \gamma - \theta_m(\gamma) = 0. \tag{5.52}$$

As discussed in section 2.3.5, this relation may not be one-to-one, so that for some values of φ more than one γ satisfies (5.52). This can be clearly seen from Figure 4.9b. The inverse dependence of γ on φ, obtained by interchanging the axes in Figure 4.9b, is plotted in Figure 5.11a for a symmetric plate and in Figure 5.11b for an non-symmetric plate. The same dependence is also obtained using the numerical solution of the phase equation (marked by circles "o" in both figures). The stationary points correspond to the propagation directions in which the waves, observed in the direction φ, are excited.

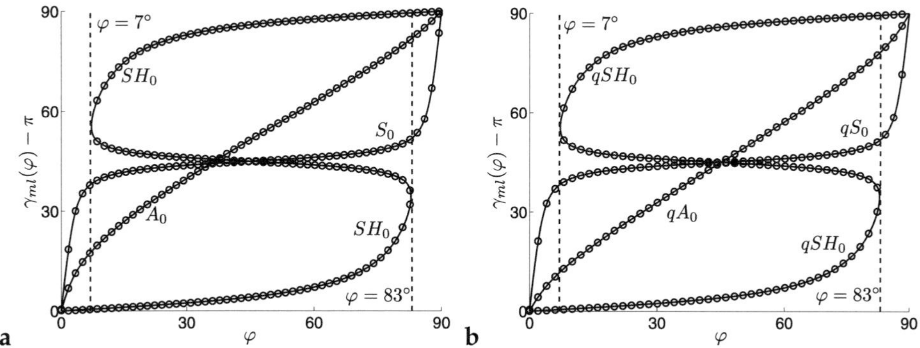

Figure 5.11: Stationary points $\gamma_{ml}(\varphi) - \pi$ of Lamb wave modes in degrees in symmetric $[0/90]_s$ (a) and non-symmetric $[0/90/0/90]$ (b) composite plates made of CFRP-T700GC/M21 (Table A.1) at frequency-thickness product 300 KHz · mm

In case of the complex pole $k_m^{\pm}(\gamma) = \operatorname{Re} k_m^{\pm}(\gamma) + \mathrm{i}\operatorname{Im} k_m^{\pm}(\gamma)$, its internal stationary points $\gamma_{mp}^{\pm}(\varphi)$ satisfy

$$\operatorname{Im}\frac{\partial P_m^{\pm}(\gamma,\varphi)}{\partial\gamma} = 0 \text{ or} \tag{5.53}$$

$$\frac{\operatorname{Im} k_{m,\gamma}'^{\pm}(\gamma)}{\operatorname{Im} k_m^{\pm}(\gamma)} = \tan(\gamma-\varphi), \quad \gamma\in(\varphi\pm\pi/2,\varphi+\pi\pm\pi/2).$$

According to the stationary phase method, the contribution of the stationary point $\gamma_{mp}^{\pm}(\varphi)$ found for the pole $k_m^{\pm}(\gamma)$ is given for $r \gg A_o$ by

$$\mathbf{G}_{mp}^{\pm}(r,\varphi,z,\omega) = \pm\frac{\mathrm{i}}{2\pi}\sqrt{\frac{2\pi}{r}}\frac{\mathbf{b}_m^{\pm}(\gamma_{mp}^{\pm}(\varphi),z,\omega)}{\sqrt{-\mathrm{i}\cdot P_{m,\gamma^2}''^{\pm}(\gamma_{mp}^{\pm}(\varphi),\varphi)}}\mathrm{e}^{\mathrm{i}rP_m^{\pm}(\gamma_{mp}^{\pm}(\varphi),\varphi)} \tag{5.54}$$

$$+ \ \mathcal{O}(r^{-3/2}),$$

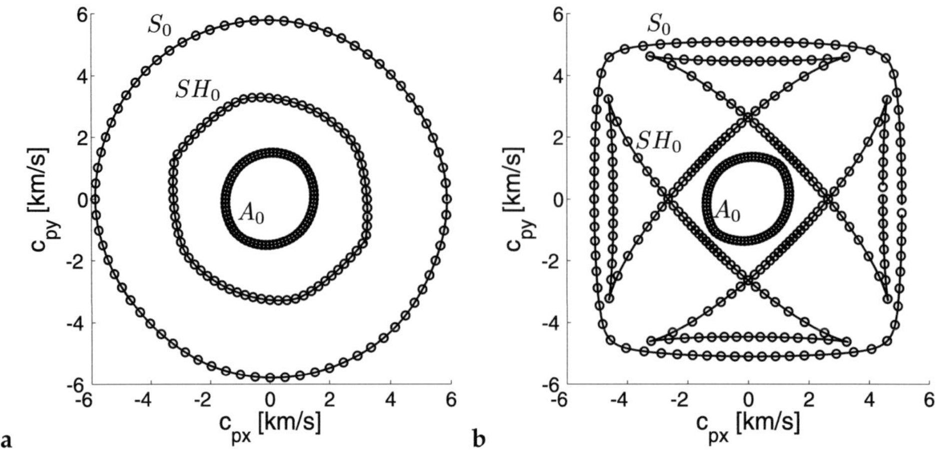

Figure 5.12: Comparison of PWS curves ("−") and the phase velocities obtained from the asymptotic expansion $\omega/P_m(\gamma_{mp}(\varphi))$ ("o") for the frequency-thickness product 800 KHz $\cdot$ mm (a) in a $[45/-45/0/90]_s$ composite plate made of AS4/3502 (Table A.1) and 500 KHz $\cdot$ mm in a $[45_6/-45_6]_s$ composite plate made of AS4/3502 (Table A.1), respectively

where

$$P''_{m,\gamma^2}(\gamma^\pm_{mp}(\varphi),\varphi) = \frac{\partial^2 P^\pm(\gamma^\pm_{mp}(\varphi),\varphi)}{\partial\gamma^2} \neq 0. \tag{5.55}$$

According to (5.54) the contribution of real poles is obtained as

$$\frac{H_{mp}(\varphi,z,\omega)}{\sqrt{r}}\,\mathrm{e}^{\mathrm{i}h_{mp}(\varphi)r} + \mathcal{O}(r^{-3/2}), \quad h_{mp}(\varphi) > 0. \tag{5.56}$$

The contribution of pure imaginary poles is

$$\frac{H_{mp}(\varphi,z,\omega)}{\sqrt{r}}\,\mathrm{e}^{-d_{mp}(\varphi)r} + \mathcal{O}(r^{-\frac{3}{2}}\mathrm{e}^{-d_{mp}(\varphi)r}) \sim \mathcal{O}(\mathrm{e}^{-r}), \quad d_{mp}(\varphi) > 0,\ h_{mp}(\varphi) > 0, \tag{5.57}$$

and, finally, the complex poles contribute as

$$\frac{H_{mp}(\varphi,z,\omega)}{\sqrt{r}}\,\mathrm{e}^{\mathrm{i}h_{mp}(\varphi)r}\,\mathrm{e}^{-d_{mp}(\varphi)r} + \mathcal{O}(r^{-3/2}\mathrm{e}^{-d_{mp}(\varphi)r}) \tag{5.58}$$

$$\sim \mathcal{O}(\mathrm{e}^{-d_{mp}(\varphi)r}) = \mathcal{O}(r^{-\infty}),$$

where $d_{mp}(\varphi) > 0$, $h_{mp}(\varphi) > 0$. For all their stationary points the contributions of imaginary and complex poles show an exponential decay with respect to r, i.e. in the far-field, their contributions are negligible. However, the contribution of each stationary point for each real pole (5.56) describes a cylindrical wave with a slowly decaying amplitude $\mathcal{O}(r^{-1/2})$ and propagating with a phase velocity given by

$$C_{mp}(\varphi) = \frac{\omega}{P_m^{\pm}(\gamma_{mp}^{\pm}(\varphi), \varphi)}. \tag{5.59}$$

The phase velocity (5.59) is independent of the type of the load. The values $C_{mp}(\varphi)$ plotted in dependence on the observation direction for all propagating wave modes and all their stationary points should agree with the PWS curves (for its calculation in Equation (2.55) the group velocity $c_g(\gamma)$ should be replaced by a phase velocity $c_p(\gamma)$), defined for an excitation by a point source. It is well illustrated in Figure 5.12, where the PWS curves are compared with curves of $C_{mp}(\varphi)$ for $[45/-45/0/90]_s$ (a) composite plate at 800 KHz · mm and for $[45_6/-45_6]_s$ (b) composite plate at 500 KHz · mm. Both plates are made of AS4/3502 (Table A.1). The main difference of Equation (5.59) in comparison to Equation (2.55) lies in the fact that Equation (5.59) defines the velocity curves explicitly whereas in Equation (2.55) the curves are defined implicitly.

Remark 5.8 *In case of an isotropic laminate, the wavenumbers are independent of γ, i.e. $k^{\pm}(\gamma) \equiv k^{\pm}$. Hence, each pole $k_m^{\pm}$ has only one stationary point $\gamma^{\pm}(\varphi) = \varphi + \pi/2 \pm \pi/2$. Moreover, $P''^{\pm}_{m,\gamma^2}(\gamma_m^{\pm}(\varphi), \varphi) = \mp k_m^{\pm}$ and Equation (5.54) in case of an excitation by a circular piezo-actuator (3.17) is derived in the far-field as follows[1]:*

$$
\begin{aligned}
G_{r,m}^{\pm}(r, \varphi, z, \omega) &= \mp \frac{\tau_0}{\sqrt{\pm 2i\pi k_m^{\pm} r}} \left(\pm i \operatorname{res} M(\alpha, z, \omega) \Big|_{\alpha=k^{\pm}} k^{\pm} \right) J_1(k^{\pm} A_o) e^{\pm i k_m^{\pm} r} \\
&\quad + \mathcal{O}(r^{-\frac{3}{2}}) \\
&= \mp \frac{i\tau_0}{\sqrt{2\pi k_m^{\pm} r}} \operatorname{res} M(\alpha, z, \omega) \Big|_{\alpha=k^{\pm}} k^{\pm} J_1(k^{\pm} A_o) e^{\pm i(k_m^{\pm} r - \pi/4)} \\
&\quad + \mathcal{O}(r^{-\frac{3}{2}}).
\end{aligned}
\tag{5.60}
$$

The equation obtained is the asymptotics of the exact analytical solution (5.38) if the asymptotics of Hankel function (B.17) for the term corresponding to the m-th pole is considered.

Note that the error in the representation (5.54), according to the method of stationary phase (B.4) is given as

$$H_{mp}^{\pm}(\varphi, z, \omega) \left(r \cdot |P''^{\pm}_{m,\gamma^2}(\gamma_{mp}^{\pm}(\varphi), \varphi)| \right)^{-3/2} + \mathcal{O}(r^{-5}), \tag{5.61}$$

i.e. it becomes to be sufficiently small only if

$$r \gg \frac{1}{|P''^{\pm}_{m,\gamma^2}(\gamma_{mp}^{\pm}(\varphi), \varphi)|}. \tag{5.62}$$

---

[1] For simplicity only the expression corresponding to $u_r(r, z, \omega)$ is provided, however, the conclusion of this remark holds also for $u_z(r, z, \omega)$.

The estimation (5.62) means that a sufficient accuracy of the computation of the displacement vector applying the asymptotic expansion (5.54) is reached in those observation directions φ faster, for which $P''^{\pm}_{m,\gamma^2}(\gamma^{\pm}_{mp}(\varphi),\varphi)$ is large. If this value is small or nearly zero, the accuracy is reached only at far distances from the excitation source.

Finally, the asymptotic expansion (AE) for the displacement vector is derived as

$$\mathbf{u}(r,\varphi,z,\omega) = \sum_{m=1}^{N_r} \sum_{p=1}^{N^{\pm}_{mp}(\varphi)} \mathbf{G}^{\pm}_{mp}(r,\varphi,z,\omega) + \mathcal{O}(r^{-3/2}), \tag{5.63}$$

where N_r is the number of real poles, and $N^{\pm}_{mp}(\varphi)$ is the number of stationary points found for the pole $k^{\pm}_m(\gamma)$ in direction φ.

The asymptotic expansion (5.63) is obtained by many authors [43, 69, 107, 109, 141] and is therefore predicted to be true in [10]. However, these authors do an estimation $\mathcal{O}(r^{-1})$, but in fact, in this thesis it is derived to be of order $\mathcal{O}(r^{-3/2})$ due to a more accurate analysis of the contributions of integrals over imaginary semi-axes in section 5.3.2.

Note that the approach, described in this section, can be applied to each term in the representation (5.37) in the coordinate system (5.36). Then, the asymptotic expansion for pointwise given load vector (P-AE) is derived in the form

$$\mathbf{u}(r,\varphi,z,\omega) \;=\; \pm\frac{i\delta_s^2}{\sqrt{2\pi}} \tag{5.64}$$

$$\times \sum_{j=1}^{N_q} \sum_{m=1}^{N_r} \sum_{p=1}^{N^{\pm}_{mp}(\widetilde{\varphi}_j)} \frac{\mathbf{K}(\alpha,\gamma^{\pm}_{mp}(\widetilde{\varphi}_j),z,\omega)\mathbf{q}_j\Big|_{\alpha=k^{\pm}_m(\gamma^{\pm}_{mp}(\widetilde{\varphi}_j))}}{\sqrt{-i\widetilde{r}_j \cdot P''^{\pm}_{m,\gamma^2}(\gamma^{\pm}_{mp}(\widetilde{\varphi}_j),\widetilde{\varphi}_j)}} \; e^{i\widetilde{r}_j P^{\pm}_m(\gamma^{\pm}_{mp}(\widetilde{\varphi}_j),\widetilde{\varphi}_j)}$$

$$+ \;\; \mathcal{O}(r^{-3/2}).$$

Note that the same asymptotic expansion is obtained also in [40]. Therefore its authors estimated that by taking the sufficient number of point sources for describing the load $\mathbf{Q}(\alpha,\gamma)$, at large distances r, where the direction $\widetilde{\varphi}_j$ becomes to be "practically parallel" to each other, the representation (5.64) turns to the asymptotic expansion (5.63).

5.5.2 Asymptotic expansion for directions near caustics

The asymptotic expansions derived in section 5.5.1, are valid if (5.55) is satisfied. On the contrary, the observation directions for which $P''^{\pm}_{m,\gamma^2}(\gamma^{\pm}_{mp}(\varphi),\varphi) = 0$ is fullfilled, correspond to caustics, and the representation (5.54) is no longer valid. In such a direction φ_c, two stationary points $\gamma^{\pm}_{mp_1}(\varphi)$ and $\gamma^{\pm}_{mp_2}(\varphi)$ join each other, i.e.

$\gamma^{\pm}_{mp_1}(\varphi) - \gamma^{\pm}_{mp_2}(\varphi) \to 0$ if $\varphi \to \varphi_c$. This situation is illustrated in Figure 5.11, where two stationary points corresponding to SH_0 and qSH_0 modes for symmetric and non-symmetric plate, respectively, become to be equal at caustics $\varphi_c \approx 7°$ and $\varphi_c \approx 83°$. In the vicinity of φ_c, where these two stationary points are nearly equal, the sum of corresponding displacements[1] is expressed for $r \to \infty$ as [29]

$$\mathbf{G}^{\pm}_{mp_1}(r,\varphi,z,\omega) + \mathbf{G}^{\pm}_{mp_2}(r,\varphi,z,\omega) = r^{-\frac{1}{3}}\,\mathrm{e}^{irL(\varphi)}V(s) \tag{5.65}$$

$$\times \left[\mathbf{b}^{\pm}_m(\gamma_{mp_2}(\varphi),z,\omega)\sqrt{\frac{2\sqrt{S(\gamma_{mp_2}(\varphi))}}{P''^{\pm}_m(\gamma_{mp_2}(\varphi))}} + \mathbf{b}^{\pm}_m(\gamma_{mp_1}(\varphi),z,\omega)\sqrt{\frac{-2\sqrt{S(\gamma_{mp_1}(\varphi))}}{P''^{\pm}_m(\gamma_{mp_1}(\varphi))}} \right]$$

$$+ \qquad\qquad\qquad\qquad \mathcal{O}(r^{-2/3}),$$

where

$$L(\varphi) = \frac{1}{2}\left[P^{\pm}_m(\gamma^{\pm}_{mp_1}(\varphi),\varphi) + P^{\pm}_m(\gamma^{\pm}_{mp_2}(\varphi),\varphi) \right], \quad s = -r^{2/3}S(\varphi), \tag{5.66}$$

$$S(\varphi) = \left[\frac{3}{4}(P^{\pm}_m(\gamma^{\pm}_{mp_2}(\varphi),\varphi) - P^{\pm}_m(\gamma^{\pm}_{mp_1}(\varphi),\varphi)) \right]^{2/3} \tag{5.67}$$

and V is the so called *Airy function*. It is assumed in (5.65) that

$$P''^{\pm}_m(\gamma^{\pm}_{mp_1}(\varphi),\varphi) < 0, \quad P''^{\pm}_m(\gamma^{\pm}_{mp_2}(\varphi),\varphi) > 0. \tag{5.68}$$

This estimation is valid when [135]

$$S(\varphi) = \mathcal{O}(r^{-\frac{2}{3}}), \quad r \to \infty, \tag{5.69}$$

or, in other words,

$$|P_m(\gamma_{m2}(\varphi),\varphi) - P_m(\gamma_{m1}(\varphi),\varphi)| = \mathcal{O}(r^{-1}), \quad r \to \infty. \tag{5.70}$$

From the left of the caustic $\varphi_c \approx 7°$ and from the right of the caustic $\varphi_c \approx 83°$, respectively (Figure 5.11), only a single stationary point is observed. The corresponding displacement field decays with $\mathcal{O}(r^{-1/2})$ (it is so called *shadow zone*), whereas from the opposite sides of caustics three stationary points exist and the displacement decays according to (5.65) with $\mathcal{O}(r^{-1/3})$. It means that at a large distance from the source, the amplitudes near the caustic are much higher compared with neighboring directions. This phenomenon causes an effect of waves (energy) focussing [20, 117]. It is also observed in numerical computations applying the spectral FEM in [138]. Note that such a theoretically obtained discontinuity of the displacement field is also observed theoretically near caustics for phonon [85] and acoustic waves [88]. Moreover, the representation based on the Airy function similar to (5.65) is obtained for elastic

---

[1]Note that in the following for the modified asymptotic expansion for directions near caustics (Equation (5.65)) the abbreviature AE-C is used.

waves in anisotropic unbounded media by Kravtsov and Orlov [68]. However, in [67] it is also noted that the wave solution near caustics is not really discontinuous due to wave difraction. For more accurate computation of the waves at far-field near caustic a generalization of the caustics to the complex caustics (considered in complex plane) similarly to the work [66] is needed. In this thesis this approach is not considered.

5.5.3 Numerical results of applying the far-field asymptotic expansion for computation of displacements

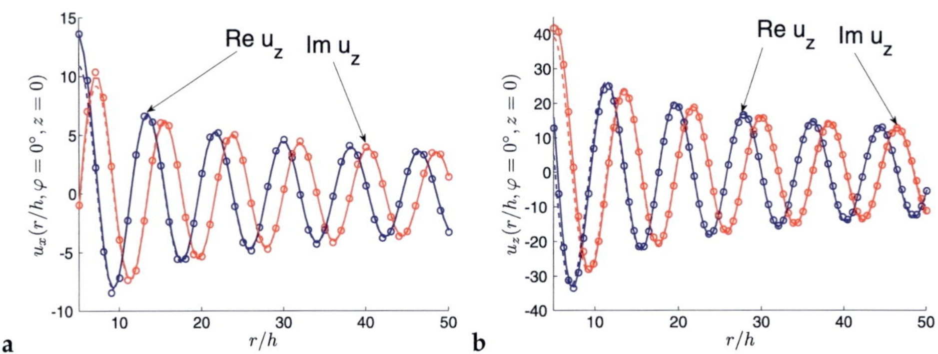

Figure 5.13: Results of the calculation of $u_x(r)$ (a) and $u_z(r)$ (b) at $z = 0$ in direction of $\varphi = 0°$ in a $[45_6/-45_6]_s$ composite plate made of AS4/3502 (Table A.1) and modelled by MLPT ($\kappa_1 = \kappa_2 = \sqrt{5/6}$). An excitation source is given by a vertical point source at $r = 0$ (3.17) acting for the frequency-thickness $f \cdot h = 100$ KHz $\cdot$ mm. Straight lines correspond to FFRIT (5.35), circles mark the results of DCI (5.7) with $R = 25$, and dashed lines are the results obtained using AE (5.63)

The asymptotic expansions given in sections 5.5.1 and 5.5.2 are applied in this section to some harmonic wave propagation problems in composite plates, the achieved results are compared with results of methods described before.

The most simplest case is a $[45_6/-45_6]_s$ composite plate modelled by MLPT ($\kappa_1 = \kappa_2 = \sqrt{5/6}$) under an excitation by a vertical point load (3.17), see Figure 5.13. There are no complex poles (except of pure imaginary poles) and the frequency-thickness $f \cdot h = 100$ KHz $\cdot$ mm is notably below the cut-off frequency-thickness of waves. The displacement vector is almost fully defined by the contribution of a bending mode A_0, which propagates as a single pulse (only one stationary point), and the corresponding wave curves are smooth, i.e. the value $P''(\gamma, \varphi)$ does not take small values and the corresponding asymptotic expansion (5.63) ("$-.$") is valid already relatively near to the excitation source. It can be observed here for both displacement components $u_x(r)$ (Figure 5.13a) and $u_z(r)$ (Figure 5.13b) by comparing with the FFRIT (5.35) ("$-$"). and the DCI (5.7) ("o").

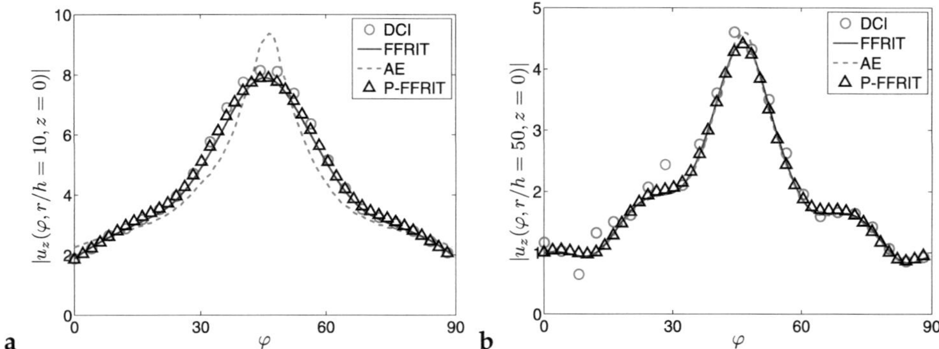

Figure 5.14: Results of the calculation of $|u_z(\varphi)|$ applying different approaches for $z = 0$, $r/h = 10$ (a) and $r/h = 50$ (b) for the frequency-thickness product $f \cdot h = 800$ KHz $\cdot$ mm in $[45/-45/0/90]_s$ composite plate made of AS4/3502 (Table A.1). An excitation source is given by a CLoVER sector (see Equation (3.22)): $\varphi_R = 22.5°$, $\varphi_R = 67.5°$, $A_i/h = 4$, $A_o/h = 5$

Another example (Figure 5.14) is given by a harmonic Lamb wave propagation in $[45/-45/0/90]_s$ composite plate under an excitation by a CLoVER sector (3.22) at $f \cdot h = 800$ KHz $\cdot$ mm, which is below the first cut-off frequency-thickness of this laminate (Table A.2 in Appendix A.9). All three propagating fundamental wave modes are taken into account. Here, due to the quasi-isotropy of all wave modes, the condition (5.62) is satisfied for relatively low values of r, however, the near-field zone is larger because of the dimensions of the actuator. In Figure 5.14a at $r/h = 10$ the innacuracy of the asymptotic expansion (5.63) is clearly observed, however at $r/h = 50$ (Figure 5.14b), it becomes to be minor in almost all directions except of directions near to $\varphi = 45°$ (the fiber direction of the upper layer in a laminate).

In some situations, the domain of applicability of the asymptotic expansion is considerably smaller, i.e. it can be applied for the displacement computation only in points far away from the excitation source. Such a situation is well illustrated by the comparison of in-plane radially directed and out-of-plane displacements in Fgiures 5.15 on the left and right sides, respectively, if computed using an asymptotic expansion ((5.63): "$-$."), far-field residue integration ((5.35): "$-$") and direct integration ((5.7): "o") in directions $\varphi = 0°$, $\varphi = 7°$ and $\varphi = 45°$. Again, only the contributions of real poles are considered and it is found that the agreement between the direct integration and far-field residue integration is well in all presented figures already for $r/h > 15$

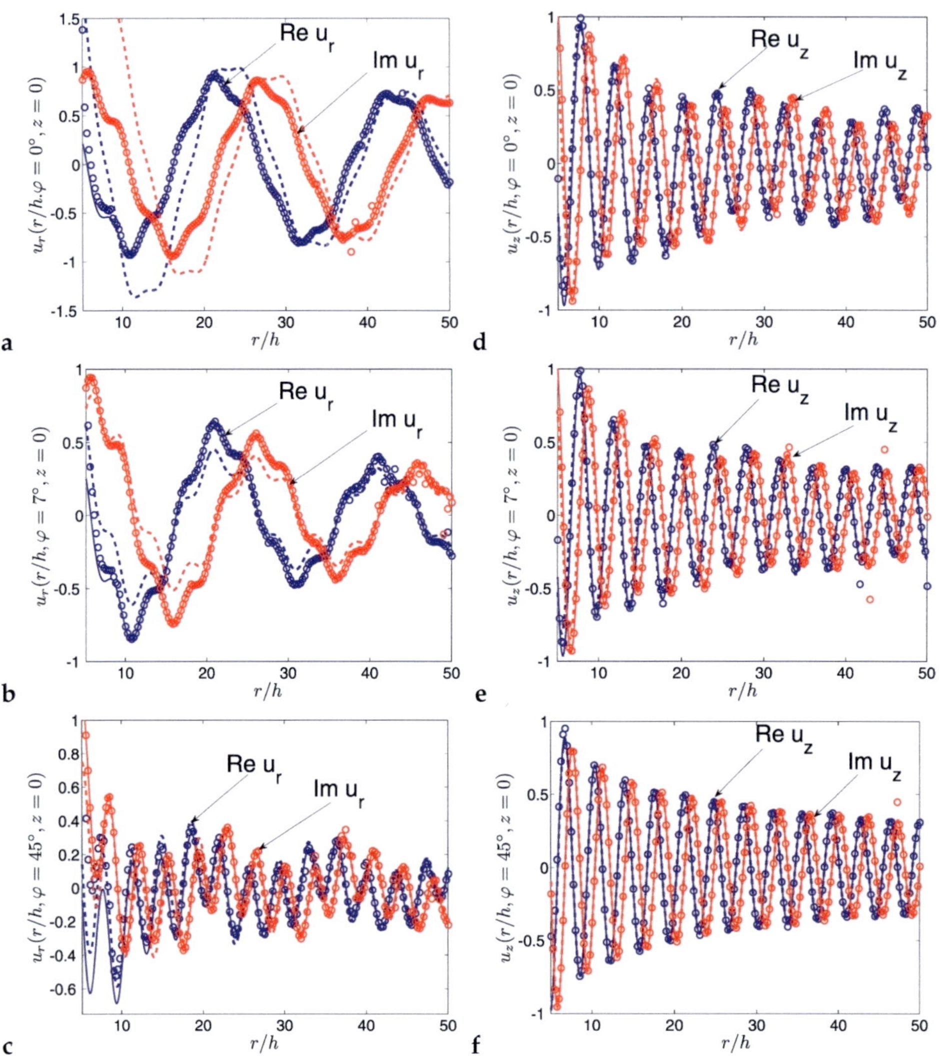

Figure 5.15: Radial in-plane (a, b and c) and out-of-plane (d, e and f) surface displacements excited by a circular piezo-actuator according to (3.17) with a radius of $A_o/h = 5$ for the frequency-thickness 300 KHz · mm in directions $\varphi = 0°$ (a and d), $\varphi = 7°$ (b and e) and $\varphi = 45°$ (c and f) in a symmetric composite plate $[0, 90]_s$ made of CFRP-T700GC/M21 (Table A.1). Comparison of results obtained using AE (5.63) ("−−"), FFRIT (5.35) ("−") and DCI (5.7) ("o")

except of the radial in-plane displacement in direction $\varphi = 45°$ (Figure 5.15c). Here, a coincidence between both approaches is reached only for $r/h > 25$. On the other hand, the asymptotic expansion agrees well for both methods for $r/h > 25$ in Figures 5.15c, d, e and f, and disagree for radial in-plane displacements in directions $\varphi = 0°$ (Figure 5.15a) and $\varphi = 7°$ (Figure 5.15b). In situations, where the asymptotic expansion gives accurate results at relatively small distances, the main contribution into the corresponding displacement components is brought by the A_0 wave mode, the velocity of which is weakly dependent on φ and the condition (5.62) is satisfied for relatively small r. In directions of $\varphi = 0°$ and $\varphi = 7°$, on the contrary, the S_0 Lamb wave mode brings the main contribution into the displacement field[1], and for this mode the condition (5.62) is satisfied only for considerably larger values of r since the second derivative of phase function is small.

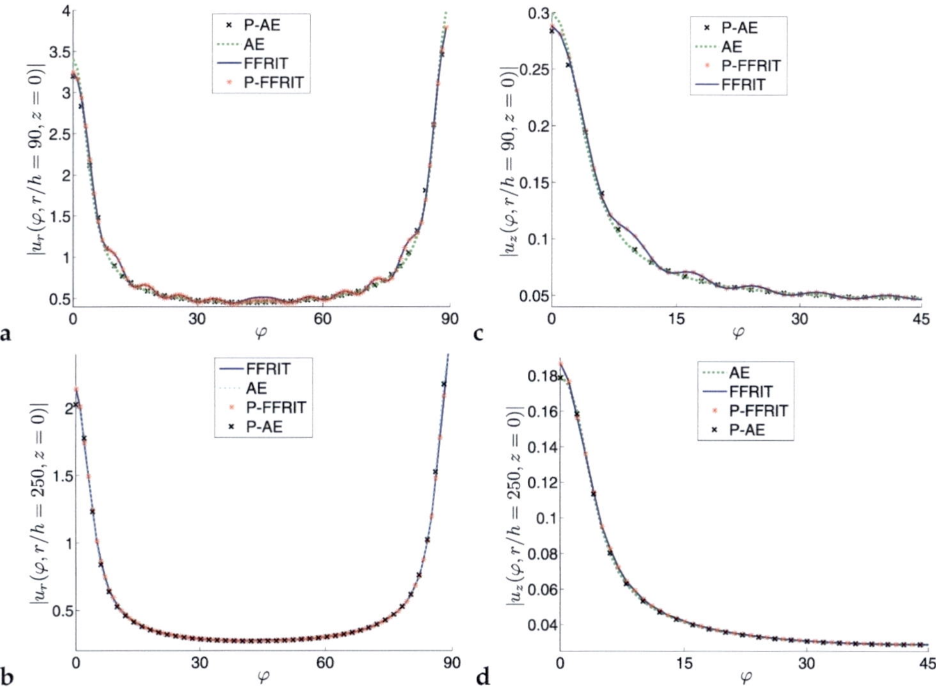

Figure 5.16: Results of Figure 5.15 but for fixed $r/h = 90$ (a and c) and $r/h = 250$ (b and d) in dependence on direction φ. Obtained applying AE (5.63) ("$--$"), P-AE ((5.64)) for $N_q = 40$ ("x"), FFRIT (5.35) ("$-$") and P-FFRIT (5.37) for $N_q = 40$ ("$*$")

---

[1]It is clearly visible because of the considerably larger wavelength in corresponding displacements.

In Figure 5.16 the values of in-plane $u_r(\varphi)$ and out-of-plane displacements $u_z(\varphi)$ are plotted. These correspond solely to the mode S_0, computed for fixed values $r/h = 90$ (Figures 5.16a and c) and $r/h = 250$ (Figures 5.16b and d) using the corresponding terms in a far-field residue integration for an exact analytical (Equation (5.35): "$-$") and pointwise (Equation (5.37): "$*$", $N_q = 40$) representations of loading function as well as using the asymptotic expansion for an exact analytical ((5.63): "$--$") and pointwise ((5.64): "x", $N_q = 40$) representations of loading function. Note that the difference between pointwise and analytical representations is insignificant at almost all directions φ. However, the difference between the use of an asymptotic expansion and the far-field residue representation is still being observable, especially for $r/h = 90$. For a large value[1] $r/h = 250$ the difference is visible only in direction $\varphi = 0°$. It let to conclude that the asymptotic expansion should be used carefully, and applied only at points, which belong to the corresponding domain of applicability of this expansion.

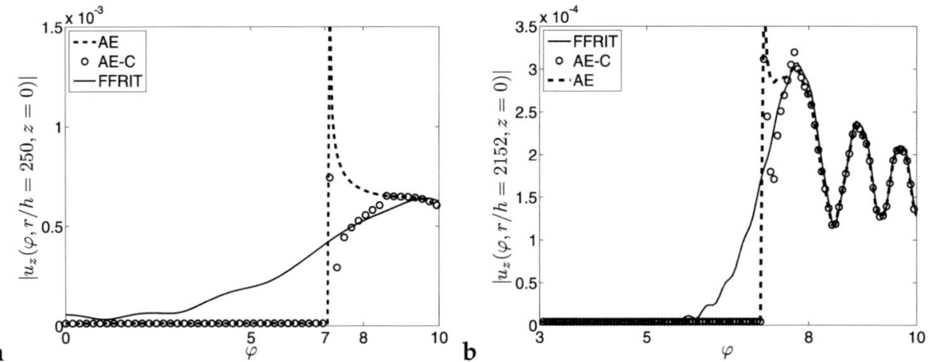

Figure 5.17: Out-of-plane displacement $|u_z(\varphi)|$ calculated for $r/h = 250$ (a) and $r/h = 2152$ (b) for $z = 0$ for the frequency-thickness 300 KHz · mm in a $[0/90]_s$ composite plate made of CFRP-T700GC/M21 (Table A.1) appying different approaches. The excitation source is given by a vertical point source at $r = 0$ (3.17)

The inaccuracies in using the asymptotic expansion occur also in directions near caustics. Figure 5.17a shows that the difference between the results obtained using the asymptotic expansion (5.63) do not agree with the results of far-field residue integration (5.35) for the mode SH_0 in directions near caustic $\varphi_c \approx 7°$ even at a distance $r/h = 250$. At the larger distance $r/h = 2152$ (Figure 5.17b) the differences are still significant even if the asymptotic expansion for directions near caustics (5.65) ("o") is used. Especially it is notable for directions from the left of caustic $\varphi_c \approx 7°$, since they propose the discontinuity of the displacements, which cannot occur for any finite value

---

[1]For a 1 mm-thick plate this distance is 25 cm.

of r as it follows from (5.35). However, at $r/h = 2152$, the accuracy of the asymptotic expansion is much higher than for $r/h = 250$.

5.6 Conclusions of the chapter

Different methods of the evaluation of $2D$-wavenumber integrals for displacements are discussed in this chapter. They are differing considerably not only in the most suitable domain of application but also in the computational costs.

The direct integration method (5.7) is slow, and in the far-field the computational costs increase rapidly - for one point on the surface of the composite plate about one minute of computational time on a standard PC is needed. On the other hand, for the evaluation of displacements in the near-field this method is most convenient since other approaches require a large number of complex poles to be taken into account or an evaluation of the integrals over the imaginary semi-axes (5.24), which leads to almost the same computational time as the direct integration. One of the main disadvantages of this approach is the impossibility of taking into account the wave structure of the solution.

After preliminarily evaluation of the stationary points and corresponding values of the phase function and its second derivative, the computational costs for the use of asymptotic expansions (5.63), (5.65) in the far-field are in the range of milliseconds on a standard PC for each point of the plate. Moreover, this approach allows to represent the displacement vector as a sum of the propagating cylindrical waves with different velocities. However, for the wave modes with a presence of caustic directions or directions of strong focussing, the domain of its applicability lies outside of the circle with a large radius $r_{\text{a.e.}}$, i.e. this approach does not work for the points inside of this considerably large circle. The value of $r_{\text{a.e.}}$ can be more than 100 times larger than the radius of the excitation domain A_0. This disadvantage is especially noticeable at higher frequencies, where the wave structure becomes to be complicated even for quasi-isotropic composites (see chapter 4). Also note that here the pointwise representation of the loading function (5.13) does not bring any advantage since the inaccuracy of the asymptotic expansion is due to the large error terms in expansions (5.63), (5.65) but not due to any inaccuracies in modelling itself.

For the domain outside the near-field, or frequently already outside of the loading domain Ω, the far-field residue integration (5.35) is convenient. This approach contains some errors brought by the assumptions made in section 5.3.2, which can be valuable in the near-field if the source is not a point source, however, it is more accurate than the use of asymptotic expansion (5.63). It also allows to analyse the contributions of single wave modes separately and is considerably faster than the direct integration (5.7) since depending on the value of r it requires on a standard PC about $1 - 5$ seconds for each value of r and φ. If in the near-field the contributions of the single wave modes are of interest, then, instead of the direct integration, the far-field residue representation for pointwise approximated loading function (5.37) can be used. However, the computational time increases proportionally to the number of nodes in a pointwise representation. But it is the only possible approach for the case when the wavenumber domain representation of the surface load is not available.

Remark 5.9 *The approaches, described in this chapter for the evaluation of displacements can be applied in the same way to the calculation of strains and stresses by considering their wavenumber domain representations.*[1]

---

[1] The representations for strains and stresses can be simply derived from this representation by applying the corresponding derivative operators.

6 Analysis of Lamb wave propagation in laminated composite plates

The algorithms for the evaluation of wavenumber integrals, presented in chapter 5, give a quite effective tool for investigation of Lamb waves excited by surface sources (e.g. piezo-electrical wafers) in laminated composites. In the following, only the far-field residue integration (5.35) and asymptotic expansion (5.63) as the two simplest and quickest ways for studying the properties of Lamb waves are considered. Since the fundamental Lamb wave modes are of main interest for practical application, their properties are studied in this chapter on various numerical examples. Amongst others, the through-thickness properties of Lamb waves, their focussing and the aspects of the transient wave propagation are discussed. Additionally, the results obtained using an integral approach applied to a transient wave propagation problem (section 3.1.2), are validated by comparing with results of conventional FEM and by comparing with experimental data.

6.1 Investigation of properties of Lamb waves

6.1.1 Through-thickness properties of Lamb waves

All techniques of the evaluation of the $2D$-wavenumber integral discussed in chapter 5, except of the direct integration (DCI), give a representation of the displacement vector as a sum of propagating Lamb wave modes. As it is stated in section 2.3.4, the Lamb wave modes in anisotropic plates are differing in their properties with respect to thickness coordinate z: there are symmetric and antisymmetric wave modes. Since in the wavenumber integral (5.1) only Green's matrix $\mathbf{K}(\alpha, \gamma, z, \omega)$ depends on z, i.e. the dependence of the displacement vector components on z is completely defined by the dependence of residues of Green's matrix $\operatorname{res}\mathbf{K}(\alpha, \gamma, z, \omega)$ for $\alpha = k_m(\gamma)$ on z and is independent of the load. In Figure 6.1 the normalized real parts of horizontal (u_x) and out-of-plane (u_z) displacements corresponding to the Lamb wave modes qS_0 (Figure 6.1a and b) and qA_0 (Figure 6.1c and d) are presented for a symmetric $[0/90]_s$ (marked by straight lines) and non-symmetric $[0/90/0/90]$ (marked by circles) composite plates made of CFRP-T700GC/M21. The excitation frequency-thickness is

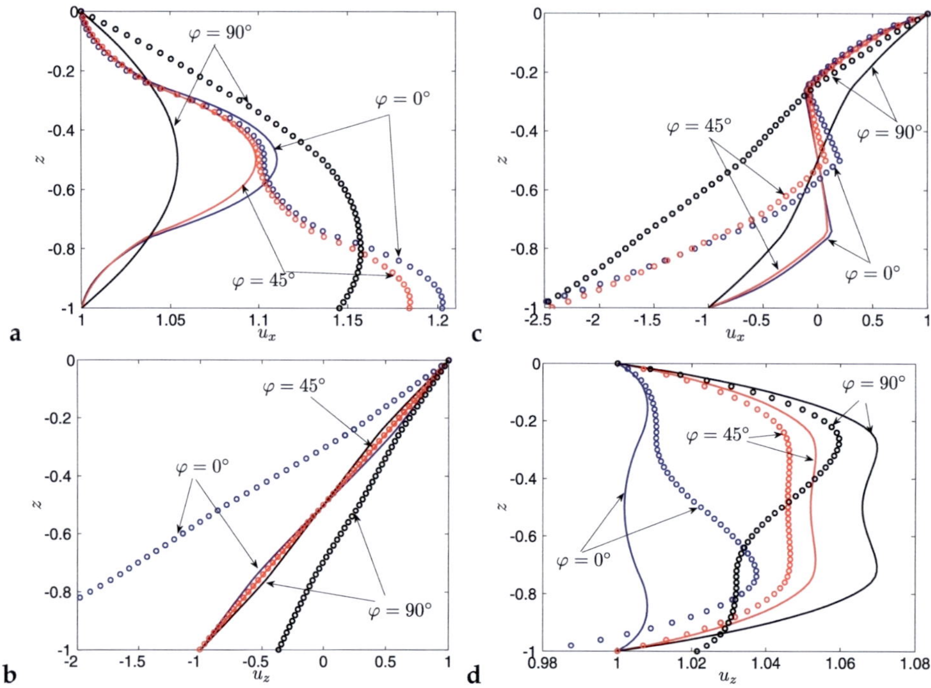

Figure 6.1: Corresponding to Lamb wave modes qS_0 (a and b) and qA_0 (c and d) normalized horizontal (a and c) and out-of-plane (b and d) components of displacement vector (real part) caused by the circular piezo-actuator (2.83) with a radius $A_o/h = 5$ for the frequency-thickness $300\,\text{KHz}\cdot\text{mm}$ in directions of $\varphi = 0°$ (black), $\varphi = 45°$ (red) and $\varphi = 90°$ (blue), computed in symmetric $[0/90]_s$ (marked with "$-$") and non-symmetric $[0/90/0/90]$ (marked with "o") composite plates with layers of CFRP-T700GC/M21

$f \cdot h = 300\,\text{KHz}\cdot\text{mm}$. Values are presented in Figure 6.1 in three observation directions: $\varphi = 0°$, $\varphi = 45°$ and $\varphi = 90°$. As it can be observed from Figure 6.1, the in-plane and out-of-plane displacements corresponding to qS_0 (or S_0) and qA_0 (or A_0) wave modes propagating in the symmetric plate $[0/90]_s$ are even or odd functions with respect to z variable, if the origin is placed at $z/h = -0.5$. This property is convenient with the classification of symmetric and antisymmetric wave modes, which is given in section 2.3.4. However, if in the $[0/90]_s$ laminated plate two bottom layers are interchanged, i.e. the non-symmetric plate $[0/90/0/90]$ is considered, the through-thickness properties of Lamb waves are changing significantly: the plots of through-thickness displacements corresponding to qS_0 and qA_0 wave modes do not display any symmetry properties of these modes with the exception of the qS_0 wave mode in the direction of $\varphi = 45°$, where this wave mode is a pure symmetric wave mode. Note that the comparison of group velocities of quasi-symmetric wave modes qS_0 and qSH_0 in both of these plates, done in chapter 4, shows no differences between their velocities.

The difference between the velocities of quasi-antisymmetric wave modes qA_0 in both plates is small, however, their through-thickness properties are significantly different.

6.1.2 Focussing of Lamb waves

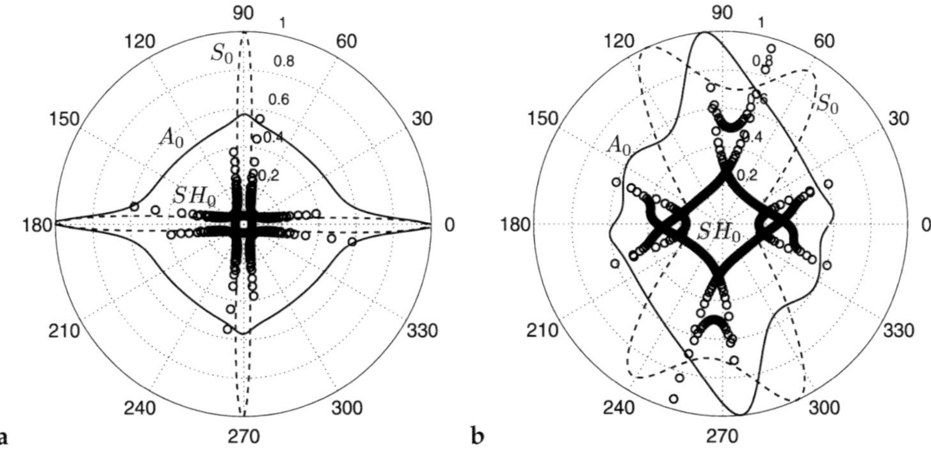

Figure 6.2: Normalized Maris factors (M_f) for fundamental Lamb wave modes A_0 (red), S_0 (blue) and SH_0 (black) for a frequency-thickness of 500 KHz · mm in a symmetric composite plate $[0/90]_s$ (a) with layers of CFRP-T700GC/M21 and in symmetric hybrid $[I90/C45/C - 45]_s$ laminated plate (b), where "I" stands for IM7-Cycom-977-3 and "C" for CFRP-T700GC/M21

In practical applications for SHM, the distribution of wave amplitudes in different observation directions for different wave modes in laminated composites is of great importance. Due to the anisotropy of the layers, the amplitudes of waves in laminated composites can exhibit a strong focussing in certain directions [20, 40]. For example, in [20] the authors observe the focussing of the S_0 wave mode in the fiber directions in a $[0/90]_s$ composite plate of CFRP-T700GC/M21 at the range of frequencies below the first cut-off frequency. As the simplest way for the prediction of the amplitudes in different directions, the authors use the Maris factor for Lamb waves, which is given by the multiplier in the asymptotic expansion (5.63) as [20, 85, 88]

$$M_{f,mp}(\varphi) = \frac{1}{\sqrt{\left| P''^{\pm}_{m,\gamma^2}(\gamma^{\pm}_{mp}(\varphi), \varphi) \right|}}.$$

(6.1)

In Figure 6.2 the normalized Maris factors of fundamental Lamb waves in a $[0/90]_s$ plate of CFRP-T700GC/M21 and a hybrid laminated plate $[I90/C45/C-45]_s$ ("I" stands for IM7-Cycom-977-3 and "C" for CFRP-T700GC/M21) are shown for the frequency-thickness of 500 KHz $\cdot$ mm. In the cross-ply plate (Figure 6.2a) the strong focussing of S_0 in the fiber directions is observed. The SH_0 wave mode is predicted to radiate preferably in caustic directions of $\varphi \approx 7°$ and $\varphi \approx 83°$, the amplitudes of A_0 are predicted to be the highest in fiber direction of the upper ply $\varphi = 0°$, $\varphi = 180°$. The angle dependence of the Maris factor for the hybrid plate (Figure 6.2b) on φ is more complicated, the A_0 mode is focussed in the direction of $\varphi \approx 97°$, the amplitudes of S_0 mode are concentrated mainly in directions of $\varphi \approx 57°$ and $\varphi \approx 113°$, whereas the SH_0 mode radiates again mainly in its caustic directions of $\varphi \approx 22°$ and $\varphi \approx 75°$.

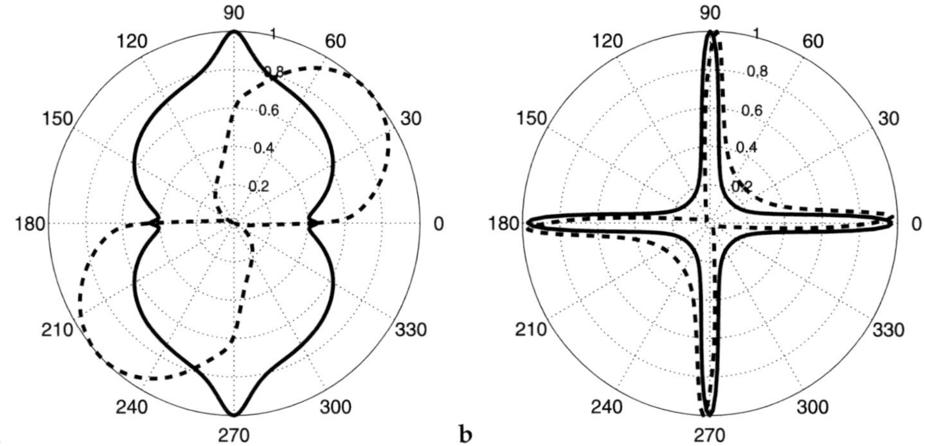

a   b

Figure 6.3: Normalized out-of-plane displacements corresponding to the fundamental Lamb wave modes A_0 (a) and S_0 (b) at a distance $r/h = 100$ in case of the excitation by the circular piezo-actuator (2.83) of radius $A_o/h = 5$ (straight lines) and the CLoVER sector (2.87) ($A_i/h = 3$, $A_o/h = 5$, $\varphi_R = \pi/8$, $\varphi_L = 3\pi/8$, dotted lines) at a frequency-thickness of 300 KHz $\cdot$ mm in a symmetric composite plate $[0/90]_s$ with layers of CFRP-T700GC/M21

The main advantage of using the Maris factor for the analysis of the directivity of wave modes is that it can be evaluated using the dispersion curves only. However, the prediction of the wave amplitudes by the Maris factor is valid under certain conditions in case of the excitation by a vertical point source since it corresponds to the multiplier in the asymptotic expansion (5.63). As such a condition, a small variation of the residues of Green's matrix with respect to γ can be chosen. However, the Maris factor does not take into account the properties of the excitation source.

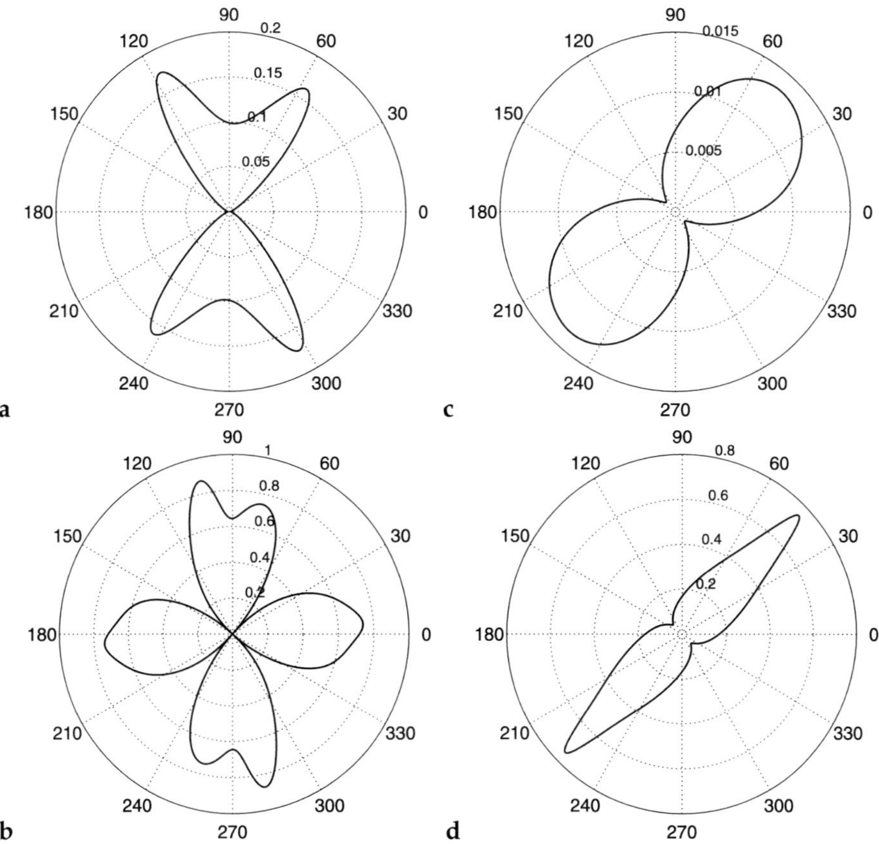

Figure 6.4: Out-of-plane displacements corresponding to fundamental Lamb wave modes S_0 (a and c) and A_0 (b and d) at a distance $r/h = 100$ in case of the excitation by a circular piezo-actuator (2.83) of radius $A_o/h = 5$ (a and b) in a hybrid $[I90/C45/C - 45]_s$ plate and by a CLoVER sector (2.87) $(A_i/h = 3, A_o/h = 5, \varphi_R = 22.5°, \varphi_L = 67.5°$, c and d) in a symmetric $[45/-45/0/90]_s$ composite plate with layers of CFRP-T700GC/M21. The harmonic excitation frequency-thickness is 500 KHz · mm

For prediction of the Lamb wave directivity in a composite plate under an excitation by a source of general type, instead of the Maris factor the distribution of the displacements with respect to the observation direction φ needs to be analysed. In Figure 6.3 the surface amplitudes corresponding to the circular excitation source (2.83) ($A_o/h = 5$, straight lines) are compared with the surface amplitudes caused by a CLoVER sector (2.87) ($A_i/h = 3$, $A_o/h = 5$, $\varphi_R = \pi/8$, $\varphi_L = 3\pi/8$, dotted lines). The A_0 (Figure 6.3a) and S_0 (Figure 6.3b) Lamb waves are simulated here applying the asymptotic expansion (5.63) for $r/h = 10$ and $f \cdot h = 300$ KHz $\cdot$ mm in $[0/90]_s$ laminated plate with layers of CFRP-T700GC/M21. The A_0 wave mode in case of an excitation by a circular piezo is mainly radiated in the direction of $\varphi = 90°$, whereas in case of the CLoVER sector it is directed in the direction of $\varphi = 45°$. In this case the use of the CLoVER sector allows to control the primary radiation direction of A_0 wave mode. Note that the Maris factor of the A_0 for this laminate predicts the main radiation in the direction of $\varphi = 0°$ (see Figure 6.2a since the Maris factors for A_0 at $f \cdot h = 300$ KHz $\cdot$ mm and at $f \cdot h = 500$ KHz $\cdot$ mm are nearly equal). In contrast to the A_0 wave mode, for the S_0 wave mode the highest amplitudes are observed in directions near to the fiber directions $\varphi = 0°$ and $\varphi = 90°$ for both sources. This means that the focussing of waves is so strong, that it cannot be controlled even by the actuator, which is designed for producing the waves in the given angular range only. Low amplitudes in this angular range cause a low sensibility of S_0 Lamb wave mode-based SHM systems.

In Figure 6.4 similar radiation diagrams for the out-of-plane displacements are plotted for S_0 (a and c) and A_0 (b and d) harmonic wave modes excited at 500 KHz $\cdot$ mm in a hybrid $[I90/C45/C - 45]_s$ laminate by a circular source (2.83) of radius $A_o/h = 5$ and in a quasi-isotropic composite $[45/ - 45/0/90]_s$ plate with CFRP-T700GC/M21 layers by a CLoVER sector (2.87). Whereas in a quasi-isotropic $[45/ - 45/0/90]_s$ laminate (Figures 6.4c and d) both waves are strongly directed in the direction given by the geometry of the CLoVER sector. In a hybrid $[I90/C45/C - 45]_s$ laminate under an axis-symmetric excitation, the amplitude curves are similar to the curves of Maris factors (Figure 6.2b) but much more complicated. Moreover, zero amplitudes in several directions ($\varphi = 0°$ for S_0 wave mode, $\varphi = 45°$ and $\varphi = 135°$ for A_0 wave mode) are observed. These correspond to the directions, where the waves from the opposite points on the actuator boundary are excited in an antiphase. From the mathematical point of view, this situation corresponds to such values of wavenumbers $k_{mp}(\varphi)$, for which the load vector of a circular piezo-actuator in a wavenumber domain (2.83) is equal to zero, i.e. $J_1(A_o k_{mp}(\varphi)) = 0$. In such directions (for a given frequency) the laservibrometer-based measurements shoud be avoided [41]. Similarly, in case of laservibrometer-based measurements such central frequencies and dimensions of piezoactuators should be avoided. Note that in Figure 6.4 the amplitudes are not normalized, and it can be observed that the out-of-plane amplitudes of A_0 wave modes are about 5 times (hybrid plate) and 20 times (quasi-isotropic plate) higher than the amplitudes of S_0 wave modes, respectively.

6.2 Analysis of energy properties of Lamb wave modes

6.2.1 Energy flow. Umov vector

Another way for the analysis of wave propagation in mechanical structures is the study of energy characteristics of the waves. In this section an energy brought into the structure by a surface source and its radiation through a structure to infinity is studied using the definitions of values of energy flow and the vector of its power density given by Umov[1] [139]. The change of the total amount of energy contained in an elastic volume V is given by the flux of the energy through its surface S [10]

$$\frac{\mathrm{d}E}{\mathrm{d}t} = \int_S e_n \, \mathrm{d}S, \qquad (6.2)$$

where $e_n(\mathbf{x}, t) = -(\partial \mathbf{u}/\partial t, \boldsymbol{\sigma}_n)$ is the density of energy flux through the surface S in the direction of its normal $\mathbf{n}$ at a point $\mathbf{x} \in S$, $\boldsymbol{\sigma}_n$ is the complex stress vector at the surface element specified by the normal $\mathbf{n}$ as $\sigma_{n,i} = \sum_{j=1}^{n} \sigma_{ij} n_j$. In (6.2) $(\mathbf{a}, \mathbf{b}) = \sum_{k=1}^{n} a_k b_k^*$ denotes the dot product of vectors, and the asterisk indicates complex conjugate values.

The energy fluxes in steady-state time-harmonic fields are given as an average change of energy amount in the domain V during one period of oscillation $T = 2\pi/\omega$, which can be reduced to the Umov-Poynting representation in terms of stresses $\boldsymbol{\sigma}_n$ and displacements $\mathbf{u}$ as

$$E^\omega = \frac{1}{T} \int_0^T \frac{\mathrm{d}E}{\mathrm{d}t} \, \mathrm{d}t = -\frac{\omega}{2} \, \mathrm{Im} \int_S (\mathbf{u}, \boldsymbol{\sigma}_n) \, \mathrm{d}S. \qquad (6.3)$$

Then, the energy brought into a composite plate by surface source $\boldsymbol{\sigma}_n(x, y, 0) = \mathbf{q}(x, y)$ acting in the domain Ω is obtained as

$$E_0^\omega = -\frac{\omega}{2} \, \mathrm{Im} \int_\Omega (\mathbf{u}(x, y, 0), \mathbf{q}(x, y)) \, \mathrm{d}x\mathrm{d}y. \qquad (6.4)$$

Using an integral approach this energy can be represented in terms of Green's matrix $\mathbf{K}(\alpha_1, \alpha_2, z, \omega)$ and the surface load vector in the frequency-wavenumber domain

$$E_0^\omega = -\frac{\omega}{8\pi^2} \, \mathrm{Im} \int_{\Gamma_1} \int_{\Gamma_2} (\mathbf{K}(\alpha_1, \alpha_2, z, \omega)\mathbf{Q}(\alpha_1, \alpha_2), \mathbf{Q}(\alpha_1^*, \alpha_2^*)) \, \mathrm{d}\alpha_1 \mathrm{d}\alpha_2. \qquad (6.5)$$

Applying the residue integration technique to the integrals in (6.5) similarly as done in chapter 5 for the representation of displacement the following formula is obtained [135]:

$$E_0^\omega = -\frac{\omega}{4\pi} \qquad (6.6)$$

$$\times \mathrm{Im} \int_0^{2\pi} \mathrm{i} \left(\sum_{m=1}^{N_r} \mathrm{res}\, \mathbf{K}(\alpha, \gamma, z, \omega) \Big|_{\alpha = k_m(\gamma)} \mathbf{Q}(k_m(\gamma), \gamma), \mathbf{Q}(k_m^*(\gamma), \gamma) \right) k_m(\gamma) \, \mathrm{d}\gamma.$$

---

[1]In the western literature it is frequently named Poynting vector.

Equation (6.6) allows to analyse the energy partition between different Lamb waves, excited by the surface load $\mathbf{q}(\mathbf{x})$.

The energy propagating from the source to infinity can be evaluated by the same way as E_0^ω. A cylinder is considered, the center of which is located at the origin and its radius is $R > A_o$ (a cylinder is taken outside of the source, which is located inside of the circle with a minimum radius A_o), and its height is equal to the thickness of the laminated plate h. Then, the energy propagating through the surface of the cylinder to infinity is given by

$$E_R^\omega(R,\omega) \;=\; -\frac{\omega R}{2}\,\mathrm{Im}\int\limits_{0}^{2\pi}\int\limits_{-h}^{0}(\mathbf{u}(R,\varphi,z),\boldsymbol{\sigma}_n(R,\varphi,z))\;\mathrm{d}\varphi\;\mathrm{d}z \tag{6.7}$$

$$=\; -\frac{\omega R}{2}\,\mathrm{Im}\int\limits_{0}^{2\pi}\int\limits_{-h}^{0}E(R,\varphi,z)\;\mathrm{d}\varphi\;\mathrm{d}z.$$

A vector of stresses $\boldsymbol{\sigma}_n$ along the surface normal $\mathbf{n} = (\cos\varphi,\sin\varphi,0)$ is calculated using the stress tensor σ_{ij} as

$$\sigma_{n,i} = \sum_{j=1}^{3}\sigma_{ij}n_j = \sigma_{i1}\cos\varphi + \sigma_{i2}\sin\varphi, \tag{6.8}$$

where the stresses σ_{ij} are connected with displacements by Hooke's law (2.1).

The dot product $(\mathbf{u}(R,\varphi,z),\boldsymbol{\sigma}_n(R,\varphi,z)) = E(R,\varphi,z)$ in the integral in (6.7) represents the density of energy flow (power density) in a point with coordinates (R,φ,z) on the surface of the cylinder along its normal. The value of power density depends on the observation angle φ, the depth z and the radius R of the cylinder. But it is evident that, if no attenuation is considered in the model, according to the law of conservation of energy, the value of total energy E_R^ω propagating to infinity through the surface of the cylinder with a radius R does not depend on its radius and is equal to the value of energy brought into the composite by a surface source

$$E_0^\omega = E_{R_1}^\omega = E_{R_2}^\omega, \quad \forall R_1 > A_o, R_2 > A_o, \tag{6.9}$$

where A_o is the radius of the minimum circle, which contains the domain Ω, on which the surface load is applied. This relation is frequently called "energy balance". However, the distribution of energy in dependence on z and φ can vary for different values of R. For example, the total through-thickness energy flow can be computed for the analysis of the total energy flow in different directions

$$E_{R,\varphi}^\omega(R,\varphi,\omega) = -\frac{\omega R}{2}\,\mathrm{Im}\int\limits_{-h}^{0}E(R,\varphi,z,\omega)\;\mathrm{d}z. \tag{6.10}$$

Remark 6.1 *In case of absence of backward propagating modes the values of energy corresponding to all wave modes are positive:* $E_{R,\varphi}^{\omega}(R,\varphi,\omega) > 0$ *and* $E_0^{\omega} > 0$.

While computing the energy flow for different values of R, φ and z, different algorithms of the displacement and the stress vector computation can be used. However, most convenient is to use the far-field residue integration (5.35) or the asymptotic expansion (5.63). Both of these approaches allow to represent the solution of the wave propagation problem as a sum of propagating wave modes, i.e. the total power density of the energy propagating from the source to infinity can be obtained as

$$E(R,\varphi,z,\omega) = \left(\sum_j \mathbf{u}_j(R,\varphi,z,\omega), \sum_k \sigma_{n,k}(R,\varphi,z,\omega) \right) \tag{6.11}$$

$$= \sum_j E_j(R,\varphi,z,\omega) + \sum_{j,k,\ j\neq k} E_{jk}(R,\varphi,z,\omega),$$

where

$$E_j(R,\varphi,z,\omega) = \left(\mathbf{u}_j(R,\varphi,z,\omega), \sigma_{n,j}(R,\varphi,z,\omega) \right), \tag{6.12}$$

and

$$E_{jk}(R,\varphi,z,\omega) = \left(\mathbf{u}_j(R,\varphi,z,\omega), \sigma_{n,k}(R,\varphi,z,\omega) \right) \tag{6.13}$$

$$+ \left(\mathbf{u}_k(R,\varphi,z,\omega), \sigma_{n,j}(R,\varphi,z,\omega) \right).$$

The values $E_j(R,\varphi,z,\omega)$ correspond to the energy, carried out solely by the i-th wave mode, i.e. these values are called *pure* energy contributions. The values $E_{jk}(R,\varphi,z,\omega)$ correspond to the energy, carried out due to the interaction of two wave modes, they are named *mixed* energy contributions. Note that the mixed energy components $E_{jk}(R,\varphi,z,\omega)$ are in general non-zero only in some directions, where the corresponding wave modes j and k are coupled.

If for the computation of the energy flow in the far-field to the excitation source the asymptotic expansion (5.63) is used, it is obtained that[1]

$$E_j(R,\varphi,z,\omega) \sim E_{jk}(R,\varphi,z,\omega) \sim \mathcal{O}(R^{-1}) \tag{6.14}$$

is fullfilled. However, in following the representation of the total energy radiated by the source (6.10) is taken into account and the following formula for the computation of the power density is used:

$$E_j^{\omega}(R,\varphi,z,\omega) = -\frac{\omega R}{2}E_j(R,\varphi,z,\omega), \quad E_{jk}^{\omega}(R,\varphi,z,\omega) = -\frac{\omega R}{2}E_{jk}(R,\varphi,z,\omega). \tag{6.15}$$

It follows immediately that

$$E_j^{\omega}(R,\varphi,z,\omega) \sim E_{jk}^{\omega}(R,\varphi,z,\omega) \sim \mathcal{O}(1), \tag{6.16}$$

---

[1]Except of the directions φ near to caustics of wave mode.

i.e. the normalized values of the power density (6.10) turn to some constant values. Note that if in the far-field the asymptotic expansion (5.63) is used for the computation of power density and there exist multiple stationary points for one of the dispersion curves, the equation of energy balance (6.9) is not exactly satisfied due to the inaccuracies of the stationary phase approach in this case.

Remark 6.2 *In case of isotropic laminates the mixed energy components are generally zero:* $E_{jk}(R, \varphi, z, \omega) \equiv 0$.

Remark 6.3 *The total mixed energy after integration with respect to z and φ should be equal to zero[1], i.e.*

$$\int_0^{2\pi} \int_{-h}^{0} E_j^{\omega}(R, \varphi, z, \omega) \, dz d\varphi = 0. \tag{6.17}$$

6.2.2 Analysis of energy characteristics

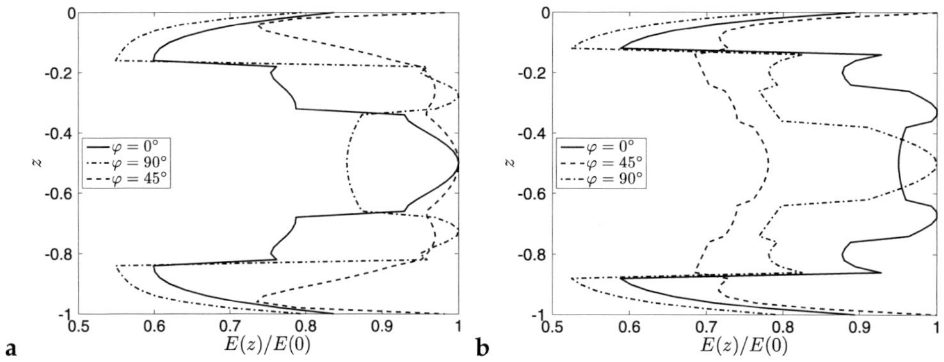

a   b

Figure 6.5: Through-thickness power density corresponding to fundamental Lamb wave mode A_0 (pure component) at a distance $r/h = 150$ in case of the excitation by the circular piezo-actuator (2.83) of radius $A_o/h = 5$ in a hybrid $[I90/C45/C - 45]_s$ plate (a) and by the CLoVER sector (2.87) $(A_i/h = 4, A_o/h = 5, \varphi_R = 22.5°, \varphi_L = 67.5°)$ in a symmetric $[45/ - 45/0/90]_s$ composite plate (b) with layers of CFRP-T700GC/M21. The harmonic excitation frequency-thickness is 500 KHz $\cdot$ mm

If the power density of the energy flow is analysed in dependece on the depth z, it can be observed that in contrast to the displacements and stresses, the power density

---

[1]In practice, some small but non-zero values of total mixed energy can occur due to numerical errors.

depends on the stresses σ_{11}, σ_{12}, σ_{22}, which can have discontinuities at the layer interfaces since only the continuity of the stress vector $(\sigma_{13}, \sigma_{23}, \sigma_{33})$ and of the displacement vector $\mathbf{u}$ is assumed (see Equation (2.8)). Note that in case of a homogeneous structure the power density is a continuous function of z [10]. The discontinuous dependence of the power density on the depth coordinate is illustrated in Figure 6.5 for the A_0 wave mode, excited in a hybrid $[I90/C45/C-45]_s$ (a) and a quasi-isotropic $[45/-45/0/90]_s$ (b) laminated plate by circular (2.83) and CLoVER sector (2.87) actuators for the frequency-thickness 500 KHz $\cdot$ mm at $r/h = 150$, respectively. Comparing the power density in different directions, it is observed that its dependence on z varies from direction to direction significantly.

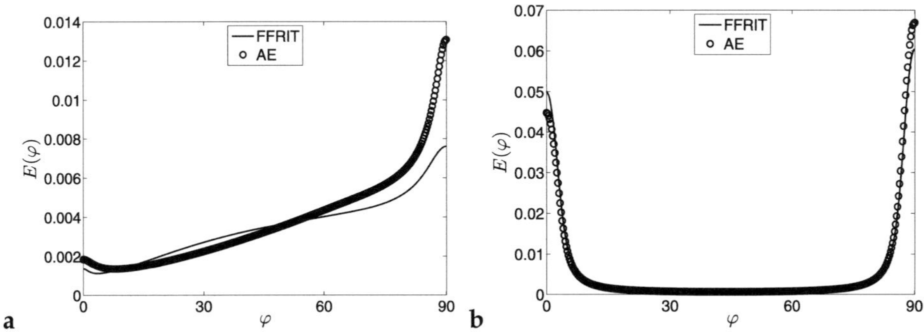

Figure 6.6: Total through-thickness energy flow $E(\varphi)$ (6.10) for wave modes A_0 (a) and S_0 (b) (pure components) in case of the actuation of waves by the circular wafer (2.83) of radius $A_o/h = 5$ at 300 KHz $\cdot$ mm in a $[0/90]_s$ plate with layers made of CFRP-T700GC/M21

After integration with respect to z (6.10), the total energy flow in the direction φ can be analysed. However, in general, the distribution of power flow over the whole range of directions $\varphi \in [0, 2\pi]$ depends on R, the distance to the origin. The application of the asymptotic expansion (5.63) gives the power density, which is independent of R since it is defined as the power density at points $R \rightarrow \infty$ (see Equation (6.16)). If instead of the asymptotic expansion the far-field residue integration (5.35) is applied, the differences between the energy distributions with respect to φ can be analysed for different values of R. As an example, the energy distributions for directions in the first quadrant $(\varphi \in [0°, 90°])$ at $R/h = 100$ are shown in Figure 6.6 for wave modes A_0 (a) and S_0 (b) propagating in a $[0/90]_s$ plate with layers of CFRP-T700GC/M21 under harmonic wave excitation by a circular wafer (2.83) $(A_o/h = 5)$ for $f \cdot h = 300$ KHz $\cdot$ mm. The differences between the values obtained applying the asymptotic expansion (Equation (5.63), marked by "o") and the far-field residue integration (Equation (5.35), straight lines) are clearly visible and the maximal values of differences are observed in directions of the fibers. Note that with increasing R, the distribution of the energy obtained by representation (5.35) turns to the one predicted by the use of asymptotic expansion (5.63), if the wave mode propagates as a single wave, i.e. only one stationary point occurs for the corresponding dispersion curve.

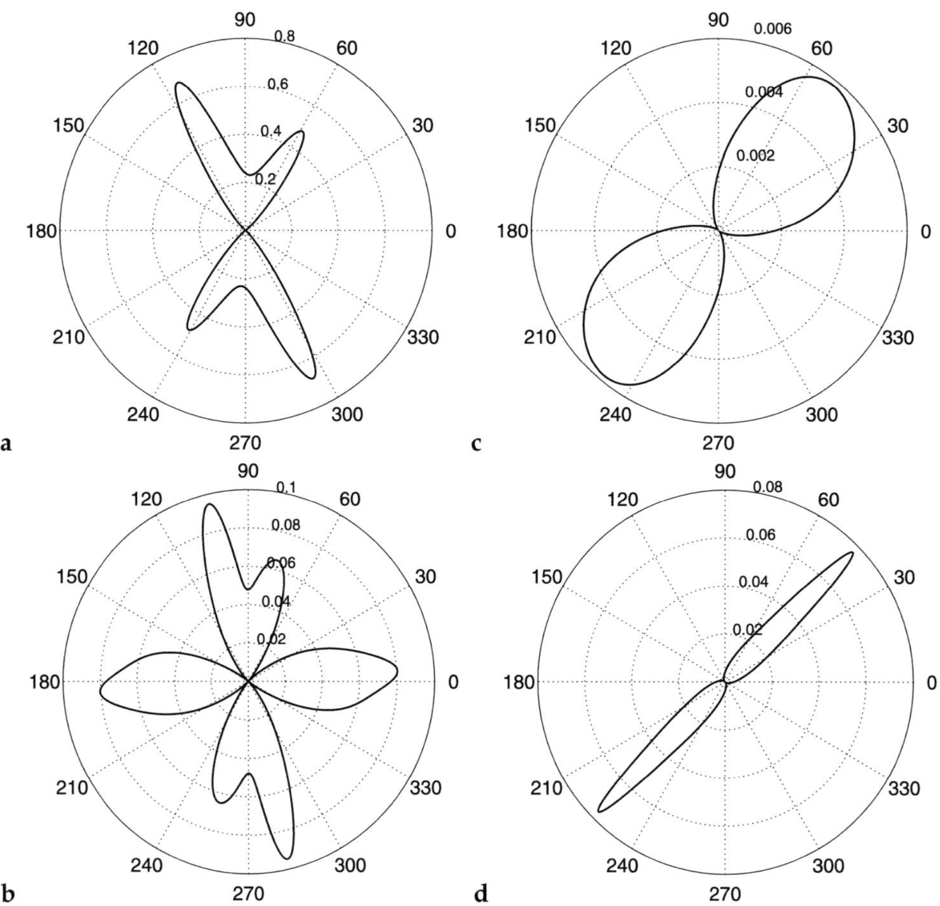

Figure 6.7: Total through-thickness power flow corresponding to fundamental Lamb wave modes S_0 (a and c) and A_0 (b and d) (both are energy components of pure modes) in dependence on φ at a distance $r/h = 150$ in case of the excitation by the circular piezo-actuator (2.83) of radius $A_o/h = 5$ in a hybrid $[I90/C45/C-45]_s$ plate (a and b) and by the CLoVER sector (2.87) $(A_i/h = 4, A_o/h = 5, \varphi_R = 22.5°, \varphi_L = 67.5°)$ in a symmetric $[45/-45/0/90]_s$ composite plate (c and d) with layers of CFRP-T700GC/M21. Harmonic excitation frequency-thickness is 500 KHz · mm

Figure 6.7 represents the distributions of energy flow of harmonic Lamb wave modes S_0 (a and c) and A_0 (b and d) excited at 500 KHz $\cdot$ mm in a hybrid $[I90/C45/C-45]_s$ laminate by a circular source (2.83) of radius $A_o/h = 5$ and in a quasi-isotropic composite $[45/-45/0/90]_s$ plate with layers of CFRP-T700GC/M21 by a CLoVER sector (2.87), respectively. It is observed that the energy radiation diagrams are quite similar to those previously discussed for the out-of-plane displacements in this case (Figure 6.4). Nevertheless, the analysis of guided Lamb waves using the energy distribution takes into account all displacement components and is more convenient for the analysis of the wave directivity since it represents the results, which are independent of the measuring procedure. For example, the method of the selective Lamb wave excitation in isotropic laminates by piezoelectric ring sources, presented in [41], is based on the analysis of the energy, radiated by the source to infinity.

6.3 Frequency spectrum of the wave propagation problem

All numerical examples considered in the previous section correspond to the solutions of the harmonic Lamb wave problem in laminated composites under the excitation by the piezoelectrical wafers. Since the algorithms for the displacement vector computation based on the residue theorem are time-efficient, a similar analysis can be done for all frequencies in the given frequency range. Since the dispersion properties of Lamb waves at high frequencies become to be complicated even in quasi-isotropic laminates (see chapter 4), in practical applications of SHM it is convenient to use the fundamental Lamb wave modes at frequencies below the first cut-off frequency. Another reason of using fundamental wave modes consists in the requirement of the use of high power electronic amplifiers for producing the waves by piezoelectric actuators at high frequencies [33]. In this section the properties of the harmonic spectrum of fundamental Lamb modes are discussed on several numerical examples.

First, the radial in-plane displacement spectrum of the sum of quasi-symmetric wave modes qS_0 and qSH_0 are calculated in a non-symmetric $[0/90/0/90]$ laminated plate (CFRP) for $z = 0$, $r/h = 150$ in dependence on the frequency-thickness product $f \cdot h$ and the observation direction φ using the model based on the elasticity theory (Equation (2.5), Figure 6.8a) and the model based on CLPT (Equation (2.36), Figure 6.8b). The waves are considered to be excited by a circular piezoelectric wafer (2.83) of the radius $A_o/h = 5$. As computational algorithm the far-field residue integration is chosen here. In the following figures, the red-colored areas correspond to high amplitudes, whereas the blue-colored areas correspond to low amplitudes. The results of both modelling approaches are well coinciding for the frequency-thicknesses below 100 KHz $\cdot$ mm. Above this frequency-thickness value, the differences between the results become to be clearly observable, i.e. the CLPT for quasi-symmetric wave modes is valid only for

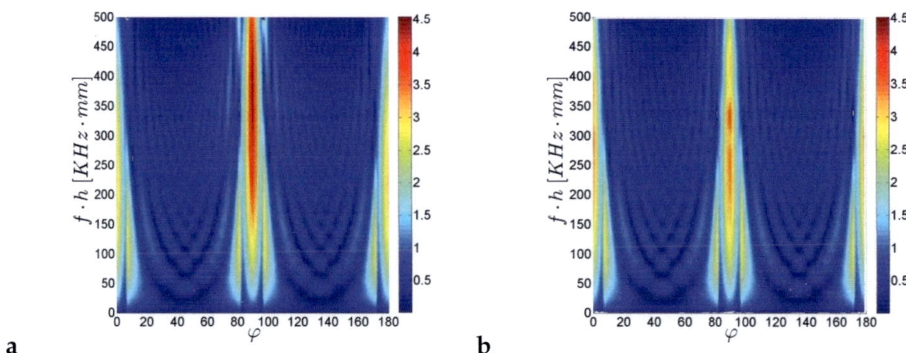

Figure 6.8: Surface radial displacements ($|u_r(\varphi, \omega)|$) for the sum of wave modes qS_0 and qSH_0 at $r/h = 150$ in case of the actuation of waves by the circular wafer (2.83) of radius $A_o/h = 5$ in a $[0/90/0/90]$ plate with layers made of CFRP-T700GC/M21. Results are calculated using the far-field residue representation (5.35) for the elasticity theory-based (a) and CLPT-based (b) models

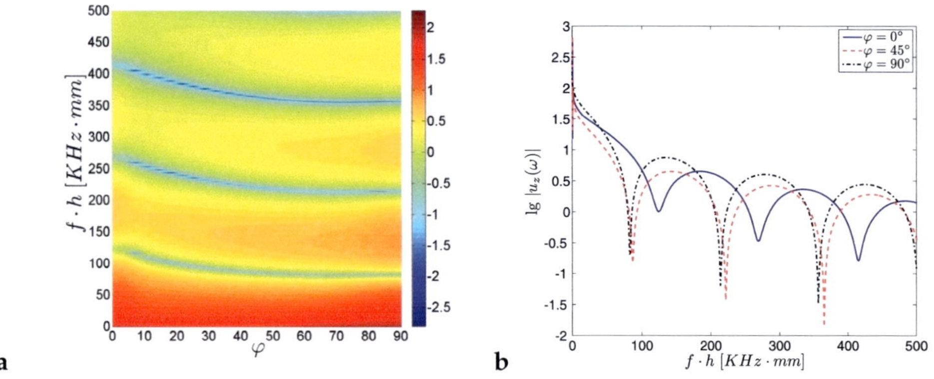

Figure 6.9: (a) Surface out-of-plane logarithmic displacements ($\lg |u_z(\varphi, \omega)|$) at $r/h = 45.7$ for the wave mode A_0 in case of the actuation of waves by the circular wafer (2.83) of radius $A_o/h = 5$ in a $[0/90]_s$ plate with layers made of CFRP-T700GC/M21. (b) Same displacements in different directions

frequency-thicknesses below 100 KHz · mm. Note that due to the non-symmetric sequence of layers in the plate, the in-plane shear tractions produced by the piezoelectric wafers also result in a non-zero out-of-plane displacement component in both linear elastic and CLPT models. If instead of a non-symmetric $[0/90/0/90]$ plate a symmetric $[0/90]_s$ plate is considered, the CLPT model is decoupled, i.e. the in-plane shear stresses produce only in-plane motions, whereas the linear elastic model is remaining to be coupled.

For the illustration of the frequency-thickness spectrum (as an analog of frequency spectrum) of the A_0 Lamb wave mode at low values of frequency-thickness, for which out-of-plane displacements are increasing as the frequency (or frequency-thickness) turns to zero, it is convenient to use logarithmic values of the displacements. In Figure 6.9a the frequency-thickness spectrum of out-of-plane logarithmical displacements of the A_0 wave mode at $r/h = 45.7$, $z = 0$ under an excitation by a circular actuator (2.83) ($A_o/h = 5$) are presented. The maximal amplitudes in different directions are comparable here, however due to the zeroes of the Bessel function $J_1(A_o k)$ in (3.17), anti-resonances of the frequency-thickness spectrum occur. Due to the dependence of wavenumbers on the observation direction, resonances and anti-resonances occur for the different directions not simultaneously. It is well observed in Figure 6.9b, where the corresponding logarithmical displacements are plotted for directions of $\varphi = 0°$, $\varphi = 45°$ and $\varphi = 90°$. The difference between the anti-resonance frequency-thicknesses is about 45 KHz · mm. It implies that in case of the resonance (or anti-resonance) excitation of waves in anisotropic laminates in one direction, in some range of other directions, the amplitudes will be lower (higher), i.e. it is nearly impossible in case of the excitation at frequency-thicknesses higher than the first anti-resonance frequency-thickness to suppress (or to spread) the wave propagation of the wave mode in all directions simultaneously.

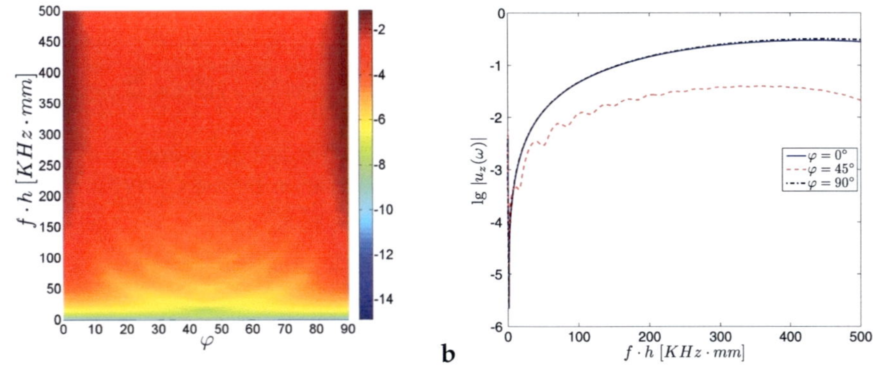

Figure 6.10: Results of Figure 6.9 but for S_0 wave mode at $r/h = 150$

For the symmetric wave mode S_0 similar computations show (Figure 6.10) that due to its higher wavelength in comparison to the A_0 mode, anti-resonances occur at frequencies higher than for the A_0 wave mode. For example, in case of an actuator of size $A_o/h = 5$, the lowest anti-resonance frequency-thickness in a $[0/90]_s$ composite is about 650 KHz · mm. Note that the strong focussing of the S_0 wave mode observed at 300 KHz · mm (Figure 6.6b) and 500 KHz · mm (Figure 6.2a) is observed in the whole low frequency range.

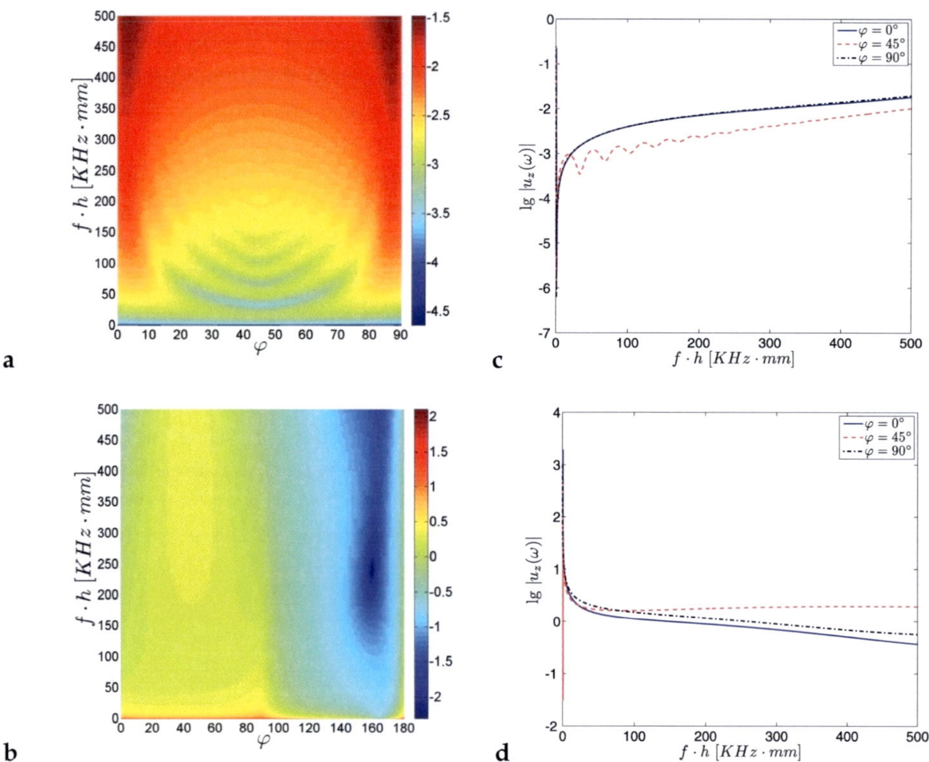

Figure 6.11: Results of Figure 6.9 and Figure 6.10 but for S_0 (a and c) at $r/h = 150$ and A_0 at $r/h = 45.7$ in case of an excitation by the CLoVER sector (2.87) ($A_i/h = 4$, $A_o/h = 5$, $\varphi_R = 22.5°$, $\varphi_L = 67.5°$)

If in the same $[0/90]_s$ plate the waves are excited by a CLoVER sector (Equation (2.87), $A_i/h = 4$, $A_o/h = 5$, $\varphi_R = 22.5°$, $\varphi_L = 67.5°$), then due to the small dimensions of the actuator the anti-resonances do not occur in a frequency-thickness range below 500 KHz · mm (see Figure 6.11). Moreover, the use of an actuator specially designed for the actuation of waves in desired sectors gives a good result for the A_0

wave mode in the whole frequency-thickness range below 500 KHz · mm. For the S_0 wave mode the maximal amplitudes are still occuring in the fiber directions for all $f \cdot h \in [25, 500]$ KHz · mm. At low frequency-thicknesses $f \cdot h \leq 25$ KHz · mm this is not true and the amplitudes of waves in all directions are of the same order.

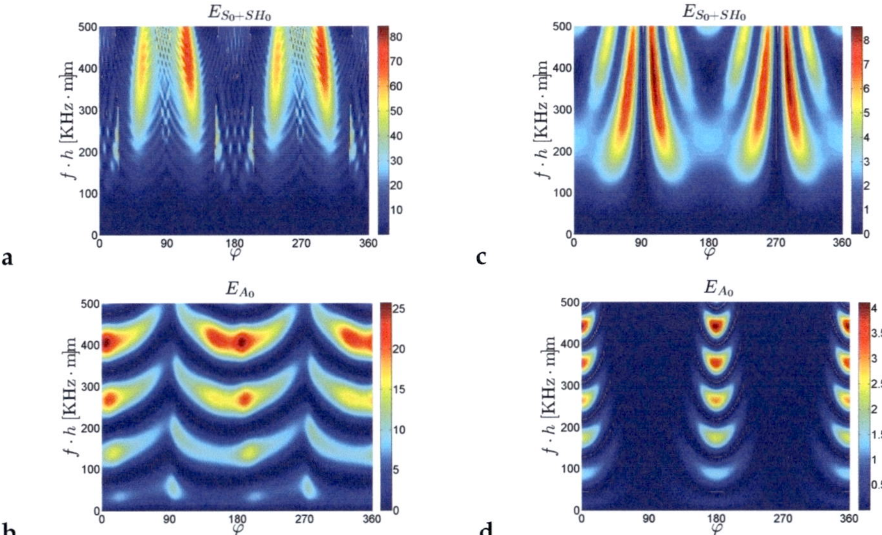

Figure 6.12: Total through-thickness energy flow in dependence on φ and ω, calculated in a hybrid $[I90/C45/C - 45]_s$ plate (a and b) under an excitation by the circular source (2.83) ($A_o/h = 5$) and in a quasi-isotropic $[0/45/ - 45/90]_{2s}$ plate (c and d) with layers of IM7-Cycom977-3 under an excitation by an MFC source ($A_1/h = 8$, $A_2/h = 2$). Values of energy flow at $r/h = 150$ for A_0 wave mode (b and d) and for the sum of S_0 and SH_0 (taking into account mixed energy components (6.13)) wave modes (a and c)

Similar to the amplitudes, the energy flow can be computed for a given frequency range. As an example, the total energy flows of the A_0 (Figure 6.12b and d) wave mode and both symmetric wave modes $S_0 + SH_0$ (Figure 6.12a and c) in dependence on φ and $f \cdot h$ are computed in a hybrid $[I90/C45/C - 45]_s$ plate excited by a circular actuator (2.83) ($A_o/h = 5$) and in a quasi-isotropic plate $[0/45/ - 45/90]_{2s}$ with layers of IM7-Cycom977-3 under an excitation by an MFC actuator ($A_1/h = 8$, $A_2/h = 2$). Whereas the directivity of the energy flow in the hybrid plate is caused by the anisotropy of the structure, the direction-dependent propagation of both wave modes in a quasi-isotropic plate is due to the use of the non-axis-symmetric source. Here an MFC actuator produces for the A_0 wave mode the power flow mostly in directions along the fiber directions in MFC ($\varphi = 0°$, $\varphi = 180°$), whereas the energy distribution of S_0 and

SH_0 is much more complicated and has many peaks in different directions at different values of frequency-thickness.

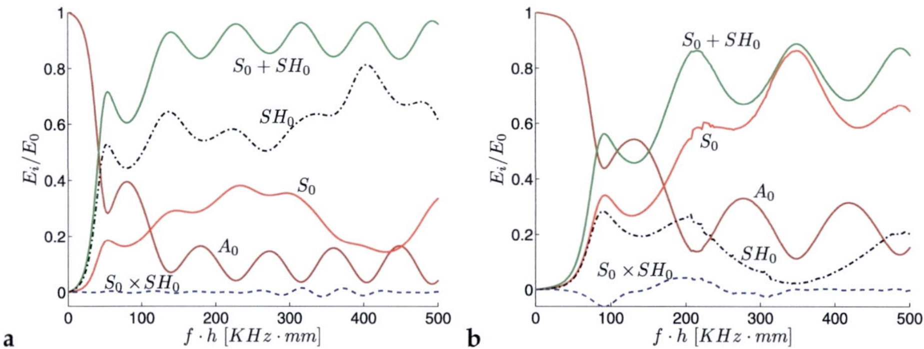

Figure 6.13: Contribution of wave modes in a total energy flow in dependence on frequency: (a) an excitation by an MFC actuator with dimensions $A_1/h = 8$, $A_2/h = 2$ in a $[0/45/-45/90]_{2s}$ composite plate of IM7-Cycom-977-3; (b) an excitation by the circular piezo actuator (2.83) of radius $A_o/h = 5$ in a hybrid $[I90/C45/C-45]_s$ composite plate

After the integration of the energy flow over all observation directions ($\varphi \in [0, 2\pi]$) and evaluating the value of the total energy E_0^ω brought into the plate by the excitation sources, the corresponding contributions of the fundamental Lamb wave modes into the whole energy flow can be analysed. For the previous example (Figure 6.12), such plots of Lamb wave mode contributions are presented in Figure 6.13. As it is observed from these figures, for low frequency-thicknesses ($f \cdot h \leq 30$ KHz $\cdot$ mm and $f \cdot h \leq 80$ KHz $\cdot$ mm for hybrid and quasi-isotropic plates, respectively) the main contribution to the total energy flow is given by the A_0 wave mode. At higher frequency-thicknesses, the main contribution of the energy is carried out by the symmetric wave modes. Note that the non-zero total contribution of the mixed energy component (Equation (6.13), $S_0 \times SH_0$) is negligible in comparison to the energy contributions of pure wave modes and caused by inaccuracies in computation of power density using the asymptotic expansion (5.63).

6.4 Solution of the transient problem

The study till this point in the thesis is focused on the analysis of the harmonic steady-state problem of wave propagation. The results obtained for the harmonic problem can be used for obtaining the solution of the more general transient problem (3.11).

6.4.1 Method based on an integral approach

For the computation of the solution of the transient problem of wave propagation according to the integral approach (3.11), it is necessary to evaluate the displacement fields for a frequency range below the frequency $\omega_{\max}$. It can be done by considering the non-uniformly distributed discrete frequencies ω_j: $0 \leq \omega_j \leq \omega_{\max}$, $j = 1, \ldots, N_\omega$, where the harmonic wave fields are computed. Then, by applying the trapezoidal rule[1], the integral (3.11) is approximated by

$$\mathbf{u}(\mathbf{x},t) \approx \frac{1}{2\pi} \operatorname{Re} \sum_{j=1}^{N_\omega-1} \left(V(\omega_j)\mathbf{u}(\mathbf{x},\omega_j)\, e^{-i\omega_j t} + V(\omega_{j+1})\mathbf{u}(\mathbf{x},\omega_{j+1})\, e^{-i\omega_{j+1} t} \right) \Delta\omega_j, \quad (6.18)$$

where $\Delta\omega_j = \omega_{j+1} - \omega_j$ and the values of $\mathbf{u}(\mathbf{x},\omega)$ are computed by one of the algorithms provided in chapter 5. Due to replacing of the improper integral (3.11) with respect to ω by an integral over the finite interval $[0, \omega_{\max}]$, it is assumed that the excitation signal $v(t)$ is sufficiently smooth, i.e. the contribution of high-frequency wave fields is negligible. However, this assumption leads to some noise in the numerical solution (6.18), which results in the small but non-zero displacements even for the timepoints before arrival of the first propagating wavefront. However, this noise does not influence the arrival time of wave packets just making the displacement curves to be non-smooth [69]. Note that the time-domain response can be obtained alternatively by the method of exponential windows [64, 78, 79, 140] or Fast Fourier Transform [80, 81].

In contrast to FEM and FD, the use of the representation (6.18) allows to evaluate the dynamic displacement at each space point and at each time point independently. In particular, it is convenient if only the displacements at a few number of points need to be computed, as it is usually required in SHM systems.

6.4.2 Comparison of the results with FEM and with experimental data

The results, obtained applying the integral approach to the transient problem are compared in this section with FEM simulations using commercial software ABAQUS [51] for a couple of test cases. The FEM models are created for symmetric wave modes by considering only the half of cross-section with respect to the thickness coordinate z with symmetric boundary conditions. Nevertheless, replacing the symmetric boundary conditions by corresponding antisymmetric boundary conditions, the results can be compared also for the case of propagation of the A_0 wave mode and higher-order antisymmetric modes [107]. Due to the simulation of Lamb waves propagation excited

---

[1]Alternatively, the data can be interpolated by splines and another more accurate cubature formula can be applied. Especially, this is necessary for (dimensionless time) timepoints $t \gg 1$.

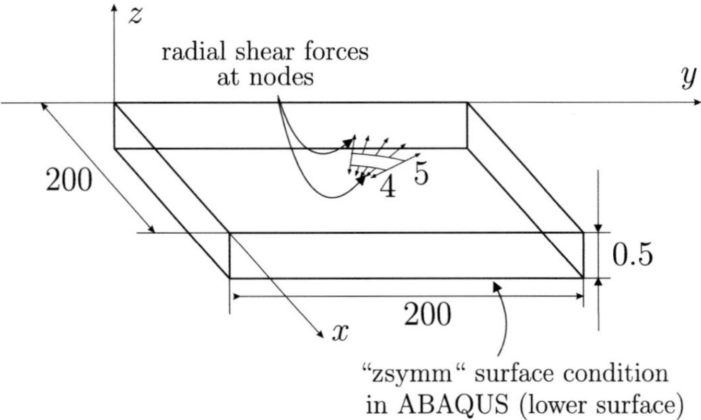

Figure 6.14: Geometry of the dimensionless FEM model for the analyis of symmetric wave motion under an excitation by the CLoVER sector (2.87)

by non-axis-symmetric sources and due to the different layouts of the laminates under consideration, the whole angular range $\varphi \in [0, 2\pi]$ is considered. However, in case of a unidirectional plate under an axis-symmetric loading (i.e. circular source), only a quarter-section of the plate needs to be considered and the computational time can be reduced. For creating the FEM mesh it is taken into account that for resolving the smallest wavelength over the frequency bandwidth at least 10 nodes per wavelength are required, and that the propagation velocities of S_0 and SH_0 wave modes are high. Finally, taking into account the dispersion properties of symmetric wave modes in a low frequency-thickness range ($f \cdot h \leq 500$ KHz $\cdot$ mm), dimensions of one brick (C3D8) element are found to be $1.3 \times 1.3 \times 0.5$ (the values are normalized with respect to the thickness h of the plate). Note that the laminated plate is modelled in ABAQUS as a composite layup, where three integration points in through-thickness directions are assigned. Such dimensions of the elements in the FEM mesh allow to simulate the wave propagation in a plate of dimensions $200 \times 200 \times 0.5$ on a standard PC. Note that the resulting FEM model has about 150000 degrees of freedom. The action of a circular-shaped piezo (2.83) of radius $A_o = 5$ is simulutad by 16 equally-distant point sources, with shear forces applied in radial direction. Another surface excitation source used for the test simulations is the CLoVER sector (see Equation (2.87)), the parameters of which are selected as $\varphi_L = 76.5°$, $\varphi_R = 22.5°$, $A_i/h = 4$, $A_o/h = 5$. The CLoVER sector is modelled in a similar way by $2 \cdot 9 = 18$ point sources located at outer and inner radii of the actuator with shear forces applied in positive and negative radial directions, respectively. The FEM simulation is carried out until the waves hit the outer boundaries of the plate. This timepoint is estimated on the basis of the propagation velocity of the S_0 wave mode. Infinite elements cannot be used to suppress reflections in these simulations, since they are not available in ABAQUS for the use of anisotropic materials [107]. Note that in the FEM calculations the time step was chosen to be small enough to accurately capture the highest excited frequency.

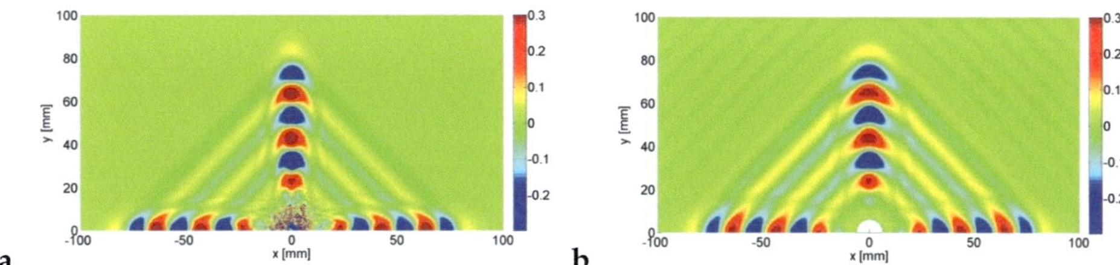

a   b

Figure 6.15: Surface out-of-plane displacements $(u_z(x,y))$ at $\bar{t} = 13.5$ for the sum of wave modes S_0 and SH_0 in case of the actuation of waves by the circular wafer (2.83) of radius $A_o/h = 5$ in a $[0/90]_s$ plate with layers made of CFRP-T700GC/M21. The excitation signal is the 3-cycles sine toneburst (2.67) with a central frequency-thickness of 300 KHz $\cdot$ mm. FEM simulation (a), semi-analytical solution (FFRIT) (b)

In Figure 6.15 the snapshots of out-of-plane surface displacements corresponding to propagating wavefronts in a $[0/90]_s$ plate of both[1] symmetric wave modes S_0 and SH_0 are presented at $\bar{t} = 13.5$, when the waves are excited by a circular actuator (2.83) applying the 3-cycles sine toneburst (2.67) with a central frequency-thickness $f_c \cdot h = 300$ KHz $\cdot$ mm. The results of the FEM simulation (Figure 6.15a) are compared with the results of applying the integral approach (FFRIT, Figure 6.15b). Except of the near-field of the excitation source, the results coincide well. Note that for the integral approach, the analytical representation for a circular source is used in contrast to the pointwise excitation of waves in the FEM model. The wavefront of the sum of both symmetric modes is illustrated here since it is impossible to decouple the symmetric wave modes S_0 and SH_0 from each other using FEM. However, applying the integral approach (e.g. FFRIT) both wave modes can be studied separately.

A similar comparison is done for the case of wave excitation by the CLoVER sector (2.87). The snapshot (Figure 6.16a) of the wavefront for S_0 and SH_0 at $\bar{t} = 16$ propagating in the first quadrant obtained by ABAQUS coincides well with similar computations using the integral approach (FFRIT, Figure 6.16b). In Figure 6.16 the waves are excited by the 3.5 Hann-modulated toneburst (2.71) with central frequency-thickness $f_c \cdot h = 300$ KHz $\cdot$ mm. Some minor differences between the wavefronts in Figure 6.16a and Figure 6.16b can be explained by the use of a pointwise representation of the load in the FEM model instead of the analytical representation of CLoVER sector in the far-field residue integration technique.

---

[1] The symmetric wave modes in FEM simulations are coupled and cannot be separated.

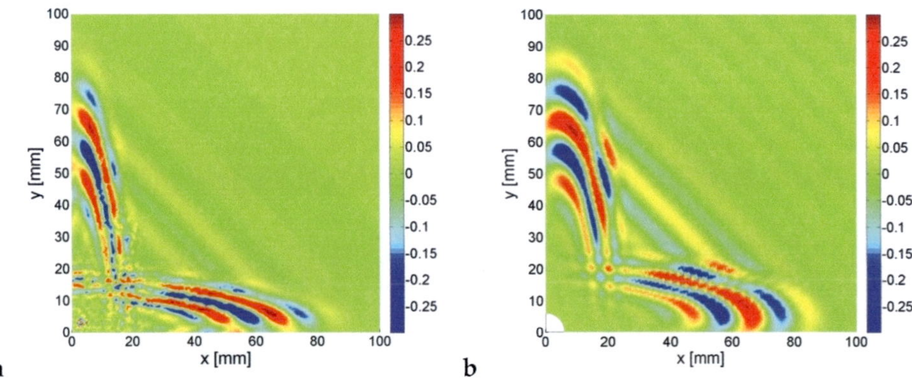

Figure 6.16: Surface out-of-plane displacements $(u_z(x,y))$ at $\bar{t} = 16$ for the sum of wave modes S_0 and SH_0 in case of the actuation of waves by the CLoVER sector (2.87) ($A_i/h = 4$, $A_o/h = 5$, $\varphi_R = 22.5°$, $\varphi_L = 67.5°$) in a $[0/90]_s$ plate with layers made of CFRP-T700GC/M21. The excitation signal is the 3.5 Hann-modulated toneburst (2.71) with a central frequency-thickness of 300 KHz · mm. FEM simulation (a), semi-analytical solution (FFRIT) (b)

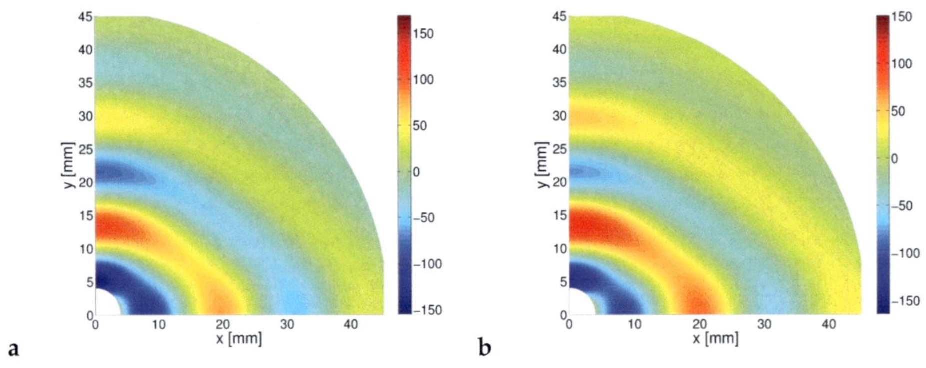

Figure 6.17: Surface out-of-plane displacements $(u_z(x,y))$ at $\bar{t} = 80$ for wave mode A_0 in case of the actuation of waves by the circular source (2.83) ($A_o/h = 5$) in a $[0/90]_s$ plate with layers made of CFRP-T700GC/M21. The excitation signal is the 3.5 Hann-modulated toneburst (2.71) with a central frequency-thickness of 25 KHz · mm. Both results are obtained applying a semi-analytical integral approach (FFRIT) for the models based on the elasticity theory (a) and MLPT (b)

In Figure 6.17 the snapshots ($\bar{t} = 80$) of the wavefront of the A_0 wave mode in a $[0/90]_s$ plate under an excitation by the circular source (2.83) ($A_o/h = 5$) with the 3.5 Hann-modulated toneburst (2.71) for $f_c \cdot h = 25$ KHz $\cdot$ mm are shown. The results on the left (Figure 6.17a) are computed using the integral approach (FFRIT) for the model based on the linear elasticity theory, the results on the right side are computed using the integral approach (FFRIT) applied to the MLPT model (2.29). It is observed from the figures that for such a low frequency-thickness, a more simple and more time-efficient MLPT model can be used for modelling the wave propagation instead of the conventional model based on elasticity theory.

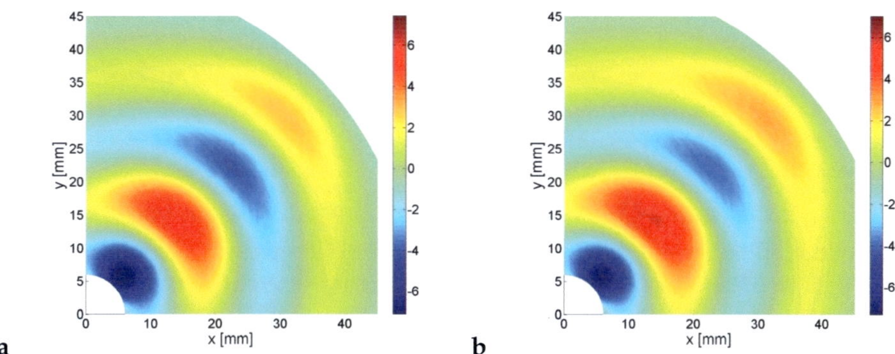

Figure 6.18: Content of Figure 6.17 but for a $[45/-45/0/90]_s$ composite plate with layers of AS4/3502 and the CLoVER excitation source (2.87) ($A_i/h = 4$, $A_o/h = 5$, $\varphi_R = 22.5°$, $\varphi_L = 67.5°$)

The similar comparison is done for a $[45/-45/0/90]_s$ laminated plate with AS4/3502 layers under an excitation by the CLoVER sector (2.87) ($A_i/h = 4$, $A_o/h = 5$, $\varphi_R = 22.5°$, $\varphi_L = 67.5°$) with the same signal as in the previous example. The results obtained by application of the integral approach (FFRIT) for the model based on the elasticity theory (Figure 6.18a) are in good agreement with the results computed using the FFRIT applied to the MLPT model (Figure 6.18b).

In Figure 6.19b the out-of-plane displacement corresponding to the first symmetric modes S_0 and SH_0 obtained using far-field residue integration technique (5.35) at $r = 10$ cm, $z = 0$ is compared for different values of φ with results [20] obtained experimentally. The waves are actuated by a PZT disk of the same size as in the theoretical model (diameter 1 cm) at a frequency of 300 KHz. The thickness of the plate under study is $h = 1$ mm. A good coincidence between numerical results and experimental data is observed. Note that the SH_0 and S_0 modes propagate at nearly equal velocities (Figure 4.8) and it is very hard to distinguish these modes in experimental data. Moreover, contribution of the shear-horizontal mode into out-of-plane displacement is

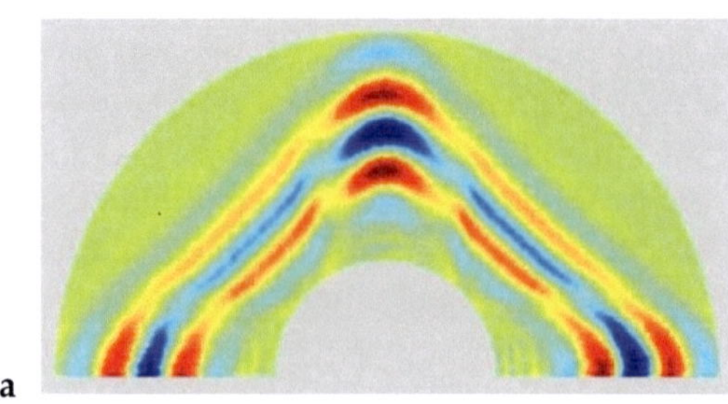
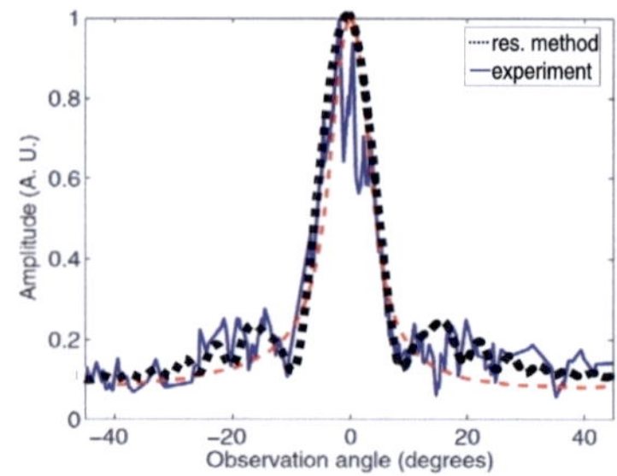

a   b

Figure 6.19: Wavefront corresponding to the sum of symmetric wave modes S_0 and SH_0 measured by a laservibrometer at $t = 22.6\ \mu s$ [20] (a). Comparison of out-of-plane displacement corresponding to S_0 and SH_0 modes due to actuation in a symmetric composite plate $[0/90]_s$ by a PZT disk of diameter 1 cm at a frequency-thickness of 300 KHz obtained using FFRIT ($--$) and experimentally ($-$) in [20] (b)

small, and out-of-plane displacement can be well predicted taking into account only S_0 as it is done in [20] using the Maris factor [85] for this mode. In Figure 6.19a the snapshot (out-of-plane displacement) of the propagating wavefront of the modes S_0 and SH_0 for the out-of-plane displacement field obtained in [20] for the experimental setup, is presented. This wavefront coincides well with the wavefronts resulting from the FEM simulation and the application of the integral approach (see Figure 6.15).

It is concluded that the results of application of semi-analytical algorithm of the computation of displacements coincide well with the FEM simulations and with at least one experimental data. It is concluded that the method described in this thesis is a promising technique in the development of a Lamb wave-based SHM system. The most critical points are lying in the assumptions done for the piezo-structure interaction (section 2.4.3) and the pure elasticity of the composite plate since the damping (or viscosity) of the layers is in general not negligible.

6.4.3 Time histories of displacements measured by laservibrometers and piezoelectric sensors

Snapshots of wavefronts at different time-points presented in the previous section are representing the process of wave propagation in composite structures quite good. Nevertheless, to get such a snapshot experimentally requires to perform many measurements using laservibrometer focussed to different points of the plate surface. However, in practice it is convenient to use only a small discrete number of points on the surface, where the measurements are performed. The time histories of the out-of-plane displacements, measured by a laservibrometer are taking the form presented

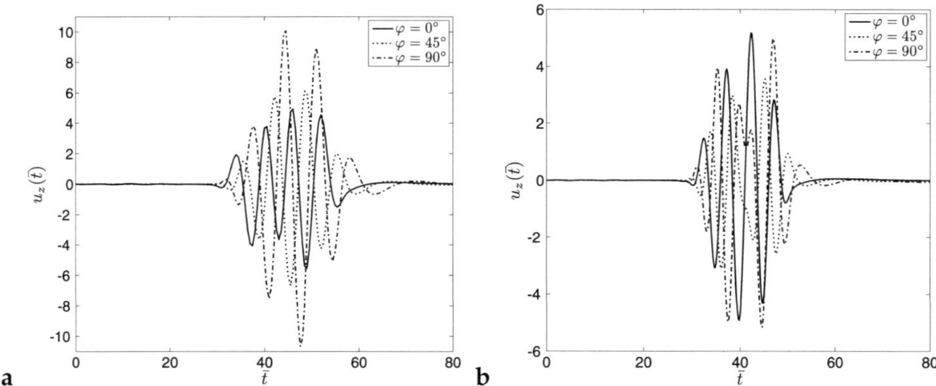

Figure 6.20: Time history of out-of-plane displacement in different directions φ for $r/h = 45.7$ for A_0 wave mode excited by the circular actuator ($A_o/h = 5$, Equation (2.83)) in a $[0/90]_s$ plate made of CFRP-T700GC/M21 with 3.5 Hann-modulated tonebursts (2.71) with central frequency-thicknesses 150 KHz · mm (a) and 215 KHz · mm(b)

in Figure 6.20. Here and hereinafter in this section the results are evaluated for a $[0/90]_s$ laminated plate. Figures 6.20a and b correspond to the time histories of out-of-plane displacement of the A_0 wave mode excited by the circular wafer (2.83) with 3.5 Hann-modulated toneburst (2.71) for central frequency-thicknesses of 150 KHz · mm (a) and 215 KHz · mm (b). They are calculated at $r/h = 45.7$, $z = 0$ in directions of $\varphi = 0°$ (straight lines), $\varphi = 45°$ (dashed lines) and $\varphi = 90°$ (dashdot lines) by use of the integral approach (FFRIT). As it follows from the corresponding frequency-thickness spectrum (Figure 6.9b), the frequency-thickness 150 KHz · mm is the resonance frequency-thickness for directions of $\varphi = 45°$ and $\varphi = 90°$. As it is predicted by the frequency-thickness spectrum, in Figure 6.20a the maximal amplitudes occur also in these directions. The time history in the direction of $\varphi = 0°$ is more complicated since the signal does not have a central frequency-thickness and is more dispersive. In Figure 6.20b due to the anti-resonance excitation in directions of $\varphi = 45°$ and $\varphi = 90°$ the corresponding time histories have two central frequency-thicknesses, i.e. the A_0 wave propagates in these directions as two pulses with different velocities. In the direction of $\varphi = 0°$ these two pulses are combining to a single pulse with the desired central frequency-thickness 215 KHz · mm.

Similar time histories at $r/h = 150$, $z = 0$ in the same directions for the S_0 wave mode excited by the circular wafer (2.83) ($A_o/h = 5$) and the CLoVER sector (2.87) ($A_i/h = 4$, $A_o/h = 5$, $\varphi_R = 22.5°$, $\varphi_L = 67.5°$) for the frequency-thickness $f \cdot h = 200$ KHz · mm are plotted in Figure 6.21a and Figure 6.21b, respectively. The corresponding frequency-thickness spectra are shown previously in Figure 6.10b and Fig-

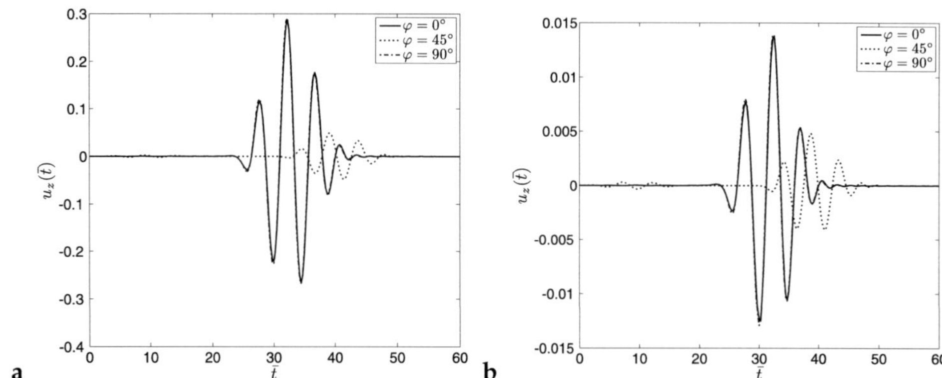

Figure 6.21: Results of Figure 6.20 but for S_0 wave mode at $r/h = 150$ under the excitation by the circular actuator ($A_o/h = 5$, Equation (2.83)) and the CLoVER sector ($A_i/h = 4$, $A_o/h = 5$, $\varphi_R = 22.5°$, $\varphi_L = 67.5°$, Equation (2.87)) using 3.5 Hann-modulated tonebursts (2.71) for central frequency-thickness 200 KHz · mm

ure 6.11c, respectively. For both excitation sources, the results in directions of $\varphi = 0°$ and $\varphi = 90°$ are nearly equal and are about 10 times (for the circular wafer) and 4 times (for the CLoVER actuator) higher than in direction $\varphi = 45°$. It means that the use of the CLoVER sector aligned along $\varphi = 45°$ allows to increase the amplitudes of propagating waves, however the waves are still focussed along the fiber directions. Strong anisotropy of the cross-ply plate under study results also in the differences in the propagation speed of the S_0 wave mode in different directions: it propagates slower in the direction of $\varphi = 45°$. Note that the amplitudes of the S_0 wave mode excited by a CLoVER sector and in case of the excitation by a circular piezoactuator cannot be directly compared with each other since different constant multipliers for the surface load (i.e. different constant voltages) are used.

The previous examples illustrate the time histories, which can be measured by the laservibrometers. For the SHM, however, it is more convenient to use for measuring the same piezoelectric elements, but now as sensors (see section 2.4.4). In this case, instead of displacements, the averaged strains are measured (2.80). The frequency-thickness spectrum of such measurements multiplied by the excitation signal is shown in Figure 6.22 for the sum of the S_0 and the SH_0 wave modes excited in a hybrid $[I90/C45/C-45]_s$ plate by the circular piezoactuator (2.83) ($A_o/h = 5$) with 3.5 Hann-modulated signal at $f_c \cdot h = 25$ KHz · mm (a) and at $f_c \cdot h = 300$ KHz · mm (b). The results are plotted as straight lines of different colours for circular sensors of radius $A_o/h = 5$, the centers of which are located at $z = 0$, $r/h = 150$ in directions of $\varphi = 0°$ (blue), $\varphi = 45°$ (green) and $\varphi = 90°$ (red). As circles corresponding results obtained applying the simplified formula (2.81) are plotted. This formula estimates the averaged

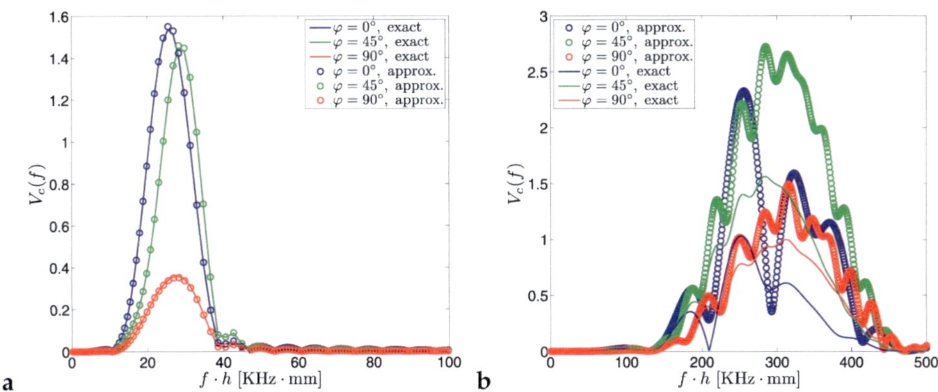

Figure 6.22: Spectra of sensor (circular sensor, $A_o/h = 5$) outputs corresponding to the sum of fundamental symmetric wave modes S_0 and SH_0 in a hybrid laminate $[I90/C45/C-45]_s$ when calculated applying exact and approximated formulas (2.80) and (2.81), respectively. Waves are excited by the circular piezo-actuator (2.83) of radius $A_o/h = 5$ using the 3.5 Hann-modulated signal (2.71) with a central frequency-thickness of $f_c \cdot h = 25$ KHz $\cdot$ mm (a) and of $f_c \cdot h = 300$ KHz $\cdot$ mm (b)

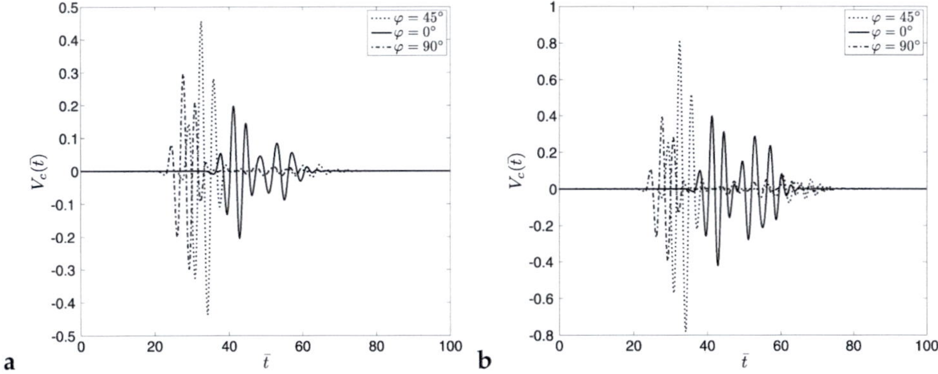

Figure 6.23: Results of Figure 6.22b but in time-domain

strains by the strains at the center of the wafer. The simplified formula is found to give accurate results only in case of a low frequency excitation (Figure 6.22a), whereas in case of the excitation with a higher center frequency-thickness $f_c \cdot h = 300$ KHz $\cdot$ mm the results are overestimated.

The differences between the use of the two formulas described are illustrated in Figure 6.23 for the corresponding time histories (excitation at $f_c \cdot h = 300$ KHz $\cdot$ mm): on the left the results of using exact[1] (Equation (2.80), Figure 6.23a) and approximated (Equation (2.81), Figure 6.23b) formulas are shown. The amplitudes predicted by the approximated formula are about two times higher than the exact values. Moreover, the structure of the signal captured by the simplified sensor (Equation (2.81), Figure 6.23b) in the direction of $\varphi = 0°$ (blue lines) differs significantly from the exact signal. Note that for the use of the exact formula due to the numerical evaluation of integrals in (2.81) the computational time is about[2] 100 times higher compared to using the simplified formula. Nevertheless, for frequency-thicknesses about $f_c \cdot h = 25$ KHz $\cdot$ mm due to the large wavelengths of symmetric wave modes it is sufficient to approximate the sensor output by the center strains of the actuator.

---

[1] Assuming an ideal bonding of piezos.

[2] The exact value of computational time depends on the number of evaluations of the function in the integral for the quadrature rule, taken for numerical evaluation.

7 Conclusion and Outlook

7.1 Conclusion

In this thesis the major aspects of the methods of the numerical simulation of time-harmonic and transient wave propagations in layered anisotropic plate-like structures based on the integral approach are examined. The results of computations are found to be in well coincidence with the results obtained applying other techniques as well as with experimental data published in [20].

The application of the integral approach to the equations of motion of elasticity theory, classical laminated plate theory and Mindlin laminated plate theory results in the solution of the problem in the transformed (i.e. wavenumber-frequency) domain as a product of Green's matrix and the load vector. Since the real poles of Green's matrix correspond to the wavenumbers of guided Lamb waves propagating in a plate, this approach can be an efficient tool for the investigation of dispersion properties of laminated composites in addition to the commonly used matrix methods. A numerical algorithm for the computation of wavenumbers in dependence on angle and frequency is presented in part 4 of this thesis. Then, its application to some numerical examples demonstrates the good coincidence of results with other results from recent publications, which are therewith approved by comparison with experimental data. The multimode structure of wave solutions, their frequency and angular dispersions as well as the influence of the stacking sequence of the laminate on the dispersive properties are investigated in this work on numerous examples, where not only wavenumbers but also the phase and group velocities of both incident and observed waves are studied. The investigation of dispersion properties of concrete composite specimens showed a highly complicated structure of high-order Lamb wave modes even in quasi-isotropic laminates. This means that energy distribution with respect to observation direction has directions of strong focussing, some wave modes are propagating as several wave packages with different velocities. These properties make difficult the application of higher-order wave modes (i.e. with short wavelengths) for purposes of SHM.

The solution of the problem in the wavenumber-frequency domain can be transformed to the time-space domain applying algorithms presented in chapters 5 and 6. The main problem here was the evaluation of the governing two-dimensional wavenumber integral, for which only the time-consuming adaptive two-dimensional numerical integration schemes or the asymptotic expressions at the far-field were known from the literature. The far-field residue integration techniques, suggested in this thesis allow to obtain the representation of the displacements as a sum of propagating Lamb waves, which is valid in both middle- and far-fields to the excitation source. Moreover, results of numerical computations presented in this thesis let to conclude that in many cases, these techniques provide quantitatively good approximation of displacements and stresses already in a near-field to the excitation source. Due to the reduction of the dimension of wavenumber integrals to one-dimensional integrals over the incident angles, this algorithm is much more time-efficient than adaptive integration techniques. Furthemore, numerical calculations showed the fitness of this approach for the calculation of displacements in directions of strong focussing since the asymptotic expansion is valid only at distances far away from the excitation source.

After the transformation of the problem into a space-frequency domain, the study of the properties of surface-excited Lamb waves in composite plates is performed for the case of steady-state actuation at different frequencies applying the techniques presented in this thesis. Then, the effect of wave focussing is investigated for various composite specimens using the displacement directivity plots and energy distribution plots. These techniques do not only allow an efficient analysis of the directivity of the excitation source at different excitation frequencies, but also provide a tool for the selective wave mode excitation and for the study of optimal design of the excitation source(s). However, in a general case[1] Lamb wave modes can be selectively generated by tuning the size of the excitation source and frequency of the excitation not in all directions simulateneously, i.e. whereas in some directions this problem will be solved nearly exact, in other directions also other wave modes will have considerable contributions. The next step consists in the direct numerical evaluation of frequency integrals, which results in the solution of the transient problem. The transient problem is solved in this work for various excitation signals. The results obtained are validated by comparing with results computed using the conventional finite element method (implemented in ABAQUS software) and by comparing with experimental results. Also by comparing with the results of calculations for the model based on elasticity theory, it is shown that the plate theories (CLPT and MLPT) give not only qualitatively but also quantitatively good results at some narrow range of low frequencies. Finally, the differences between measuring of displacements (strains) by laser-vibrometers and piezoelectrical wafers used as sensors are shown based on several numerical examples. Here a more accurate integral model for the waves excited at middle and high frequencies is found to be more appropriate.

---

[1]Instead of frequency ranges with quasi-isotropic behaviour of waves.

Summarizing the above conclusions, the main results of this work are listed as

1. An algorithm based on the application of the integral approach is developed for the evaluation of the frequency-wavenumber domain solution of 3D wave propagation problem in laminated composites consisting of the layers of arbitrary elastic anisotropy considering the equations of motion of elasticity theory. This solution is obtained as the product of Green's matrix and the representation of the surface source in the transformed domain. The algorithm of Green's matrix evaluation is proved to be stable since no growing exponents appear on all stages of the computation.

2. A stable numerical algorithm for the calculation of Lamb wave dispersion curves of incident and observed waves in laminated composite plates is developed in this thesis. It uses the dispersion relations obtained from the properties of Green's matrix in the transformed domain. The resulting dispersion curves are validated by comparing with the results published in literature and by comparing with the dispersion curves for plates by CLPT and MLPT.

3. A time-efficient algorithm of the evaluation of $2D$-wavenumber integrals for points located in middle- and far-fields, namely the far-field residue integration technique is developed. Its efficiency and high accuracy are proven for the numerical examples by comparing with accurate but time-consuming adaptive integration schemes and with an asymptotic expansion. The accuracy of far-field residue integration technique is studied for the isotropic laminates analytically whereas for anisotropic laminates the approximate formula is obtained.

4. Formulas for the calculation of the $2D$-wavenumber integral using the far-field residue integration technique and the asymptotic expansion are provided for the case of modelling of the surface source by several point sources.

5. The formula for the computation of the $2D$-wavenumber integral for directions near caustics is obtained in this thesis applying the Airy function.

6. All the methods developed in this work are implemented on PC and tested on various numerical examples considering the harmonic steady-state as well as transient wave propagation problems. The results are validated by comparing with results of FEM simulations and data obtained experimentally.

7. Multiple simulations done for various excitation sources and laminated plates show the applicability of this approach for optimization of the wave fields for structural health monitoring of the plate-like structures (i.e. the choice of a more suitable wave mode, the choice of optimal number, form and size of the actuation sources and their positions, the choice of the optimal excitation frequency). This optimization can be done by investigating the energy properties, directivity of displacements and values of sensor outputs as it is shown on numerical examples in this work.

7.2 Outlook

The study presented in this thesis does not yet solve all questions for the creation of a stable and effective SHM system based on elastic waves. However, the algorithms given in this work provide a quite good tool for development of other algorithms solving these questions. The main problems for future studies are listed as follows:

- Extension of the integral approach on the material models given by the higher-order plate theories. This step can allow to simplify and to accelerate the computation of displacements at frequencies in the mostly used frequency range below the cut-off frequencies of higher-order wave modes.

- Investigation of potential of Lamb waves at higher frequencies including higher-order wave modes.

- Tuning for higher efficiency: the development of techniques for selective wave mode excitation, excitation of waves with a given directivity (by one or several actuators similarly to phased arrays) and the optimization of power required for wave actuation.

- Study of the influence of damping on wave propagation in laminated fibre-reinforced composites since they are known to attenuate the waves due to the viscosity of the matrix in which the fibers are placed.

- Another issue is the investigation of the contact between the host structure (laminated plate) and the piezoelectric actuators bonded on its surface in case of using both as actuators and as sensors. Since from previous works it is known that the pin-force model provides accurate results only in the low frequency range and does not take into account the interaction of piezoelectrical wafers with propagating waves, it is important to study the solution of the related contact problems.

- An obligatory step for the development of Lamb wave based SHM systems consists in the simulation of wave interaction with damages and in the development of the effective algorithms of damage identification.

Appendix

A Elastic properties of composite materials

The properties of pure elastic materials summarized in this section are provided in various books on modelling of anisotropic materials [56, 74, 113]

A.1 Voigt notation

The vectors of strain $\boldsymbol{\varepsilon}$ and stress $\boldsymbol{\sigma}$ used in a representation (2.2) are given in Voigt notation as

$$
\begin{aligned}
\varepsilon_1 &= \varepsilon_{xx} = u_{x,x}, \quad \varepsilon_2 = \varepsilon_{yy} = u_{y,y}, \quad \varepsilon_3 = \varepsilon_{zz} = u_{z,z}, \quad \varepsilon_4 = \gamma_{zy} = 2\varepsilon_{zy} = u_{z,y} + u_{y,z}, \\
\varepsilon_5 &= \gamma_{zx} = 2\varepsilon_{zx} = u_{z,x} + u_{x,z}, \quad \varepsilon_6 = \gamma_{xy} = 2\varepsilon_{xy} = u_{x,y} + u_{y,x}, \\
\sigma_1 &= \sigma_x, \quad \sigma_2 = \sigma_y, \quad \sigma_3 = \sigma_z, \quad \sigma_4 = \sigma_{zy}, \quad \sigma_5 = \sigma_{zx}, \quad \sigma_6 = \sigma_{xy}.
\end{aligned}
\tag{A.1}
$$

The stiffness matrix $\mathbf{C}$ in (2.2) is obtained from the stiffness tensor $\mathcal{C}$

$$
\{C_{ij}\} =
\begin{pmatrix}
C_{1111} & C_{1122} & C_{1133} & C_{1123} & C_{1131} & C_{1112} \\
C_{2211} & C_{2222} & C_{2233} & C_{2223} & C_{2231} & C_{2212} \\
C_{3311} & C_{3322} & C_{3333} & C_{3323} & C_{3331} & C_{3312} \\
C_{2311} & C_{2322} & C_{2333} & C_{2323} & C_{2331} & C_{2312} \\
C_{3111} & C_{3122} & C_{3133} & C_{3123} & C_{3131} & C_{3112} \\
C_{1211} & C_{1222} & C_{1233} & C_{1223} & C_{1231} & C_{1212}
\end{pmatrix}
\tag{A.2}
$$

The relation between the stress and strain vectors is frequently presented in an inversed form

$$
\varepsilon_i = S_{ij}\sigma_j, \quad i = 1,\ldots,6,
\tag{A.3}
$$

where $\mathbf{S} = \mathbf{C}^{-1}$ is the so-called compliance matrix. It is a historic absurdity that stiffness is denoted by $\mathbf{C}$ and compliance by $\mathbf{S}$ [56]!

A.2 Transformation of coordinates for the stiffness matrix

For the transformation matrix between two coordinate systems given by direction cosines of the angles between the old and new coordinate axes as

$$
\mathbf{a} = \begin{pmatrix} l_1 & m_1 & n_1 \\ l_2 & m_2 & n_2 \\ l_3 & m_3 & n_3 \end{pmatrix},
\tag{A.4}
$$

when the stiffness matrix coordinates in a new coordinate system (2.3) are given in Voigt notation by

$$
C'_{ij} = q_{ik} q_{jl} C_{kl},
\tag{A.5}
$$

where the 6×6 matrix q is defined [74] as

$$
q = \begin{pmatrix}
l_1^2 & m_1^2 & n_1^2 & 2m_1 n_1 & 2n_1 l_1 & 2l_1 m_1 \\
l_2^2 & m_2^2 & n_2^2 & 2m_2 n_2 & 2n_2 l_2 & 2l_2 m_2 \\
l_3^2 & m_3^2 & n_3^2 & 2m_3 n_3 & 2n_3 l_3 & 2l_3 m_3 \\
l_2 l_3 & m_2 m_3 & n_2 n_3 & m_2 n_3 + m_3 n_2 & n_2 l_3 + n_3 l_2 & l_2 m_3 + l_3 m_2 \\
l_3 l_1 & m_3 m_1 & n_3 n_1 & m_3 n_1 + m_1 n_3 & n_3 l_1 + n_1 l_3 & l_3 m_1 + l_1 m_3 \\
l_1 l_2 & m_1 m_2 & n_1 n_2 & m_1 n_2 + m_2 n_1 & n_1 l_2 + n_2 l_1 & l_1 m_2 + l_2 m_1
\end{pmatrix}.
\tag{A.6}
$$

A.3 The types of symmetries of materials

When the stiffness matrix (A.2) is fully occupied, it describes the *triclinic*, i.e. fully anisotropic material. It is the most general case of the anisotropy because there are no planes of symmetry for the material properties. However, composite materials usually involve some symmetries of their properties. If the material has one plane of material property symmetry, e.g. $z = 0$, it is called *monoclinic* and characterized by 13 independent stiffness constants

$$
\begin{pmatrix}
C_{11} & C_{12} & C_{13} & 0 & 0 & C_{16} \\
C_{12} & C_{22} & C_{23} & 0 & 0 & C_{26} \\
C_{13} & C_{23} & C_{33} & 0 & 0 & C_{36} \\
0 & 0 & 0 & C_{44} & C_{45} & 0 \\
0 & 0 & 0 & C_{45} & C_{55} & 0 \\
C_{16} & C_{26} & C_{36} & 0 & 0 & C_{66}
\end{pmatrix}.
\tag{A.7}
$$

In case of two orthogonal planes of material property symmetry, the symmetry exists relative to a third mutually orthogonal plane. The stiffness matrix in coordinates aligned with *principal material directions*[1] are

$$
\begin{pmatrix}
C_{11} & C_{12} & C_{13} & 0 & 0 & 0 \\
C_{12} & C_{22} & C_{23} & 0 & 0 & 0 \\
C_{13} & C_{23} & C_{33} & 0 & 0 & 0 \\
0 & 0 & 0 & C_{44} & 0 & 0 \\
0 & 0 & 0 & 0 & C_{55} & 0 \\
0 & 0 & 0 & 0 & 0 & C_{66}
\end{pmatrix} .
\tag{A.8}
$$

Composites with fully unidirectional reinforcement are approximately *transversely isotropic* materials (i.e. if all fibers are oriented in x direction when directions y and z are equal). A transversely isotropic material is symmetric with respect to a rotation about an axis of symmetry, e.g. the x-axis. The stiffness matrix has 5 independent constants:

$$
\begin{pmatrix}
C_{11} & C_{12} & C_{12} & 0 & 0 & 0 \\
C_{12} & C_{22} & C_{23} & 0 & 0 & 0 \\
C_{12} & C_{23} & C_{22} & 0 & 0 & 0 \\
0 & 0 & 0 & (C_{22} - C_{23})/2 & 0 & 0 \\
0 & 0 & 0 & 0 & C_{66} & 0 \\
0 & 0 & 0 & 0 & 0 & C_{66}
\end{pmatrix} .
\tag{A.9}
$$

If all directions in the material are equal, then the material is termed *isotropic* and its stiffness matrix is given only by 2 independent constants

$$
\begin{pmatrix}
C_{11} & C_{12} & C_{12} & 0 & 0 & 0 \\
C_{12} & C_{11} & C_{12} & 0 & 0 & 0 \\
C_{12} & C_{12} & C_{11} & 0 & 0 & 0 \\
0 & 0 & 0 & (C_{11} - C_{12})/2 & 0 & 0 \\
0 & 0 & 0 & 0 & (C_{11} - C_{12})/2 & 0 \\
0 & 0 & 0 & 0 & 0 & (C_{11} - C_{12})/2
\end{pmatrix} .
\tag{A.10}
$$

A.4 Engineering notation for elastic constants

Engineering constants are more appropriate for industrial applications because they have an obvious physical interpretation and can be measured in simple tests such as uniaxial tension or pure shear tests. The compliance matrix for an *orthotropic* material in terms of engineering constants is given as

$$
\mathbf{S} = \mathbf{C}^{-1} =
\begin{pmatrix}
1/E_x & -\nu_{yx}/E_y & -\nu_{zx}/E_z & 0 & 0 & 0 \\
-\nu_{xy}/E_x & 1/E_y & -\nu_{zy}/E_z & 0 & 0 & 0 \\
-\nu_{xz}/E_x & -\nu_{yz}/E_y & 1/E_z & 0 & 0 & 0 \\
0 & 0 & 0 & 1/G_{yz} & 0 & 0 \\
0 & 0 & 0 & 0 & 1/G_{xz} & 0 \\
0 & 0 & 0 & 0 & 0 & 1/G_{xy}
\end{pmatrix} ,
\tag{A.11}
$$

---

[1]Principal material directions are directions that are parallel to the intersections of the three orthogonal planes of material property symmetry.

where

- E_x = longitudinal modulus of elasticity,

- E_y = transverse modulus of elasticity,

- E_z = through-thickness modulus of elasticity,

- G_{xy} = longitudinal in-plane shear modulus,

- G_{xz} = longitudinal through-thickness shear modulus,

- G_{yz} = transverse through-thickness shear modulus,

- v_{xy} = major Poisson's ratio,

- v_{xz} = minor Poisson's ratio,

- v_{yz} = transverse Poisson's ratio.

Note that Poisson's ratios are not symmetric, i.e. $v_{xy} \neq v_{yx}$.

For *transversely-isotropic* material, the following relations between engineering constants are satisfied:

$$E_y = E_z, \quad v_{xy} = v_{xz}, \quad v_{yx} = v_{zx}, \quad G_{xy} = G_{xz}, \quad G_{yz} = E_y/(2(1+v_{yz})). \quad (A.12)$$

The engineering constants for the *isotropic* case are presented by two independent constants, the well-known Young's modulus E and Poisson's ratio v:

$$E = E_i, \quad v = v_{ij}, \quad G = G_{ij} = E/(2(1+v)), \quad i,j = 1,2,3. \quad (A.13)$$

A.5 Elastodynamic equations in matrix form

The matrices occuring in the elastodynamic equations (2.7) are given as

$$A^{(01)} = \begin{pmatrix} C_{11} & C_{16} & C_{15} \\ C_{61} & C_{66} & C_{65} \\ C_{51} & C_{56} & C_{55} \end{pmatrix}, \qquad (A.14)$$

$$A^{(02)} = \begin{pmatrix} C_{66} & C_{62} & C_{64} \\ C_{26} & C_{22} & C_{24} \\ C_{46} & C_{42} & C_{44} \end{pmatrix}, \qquad (A.15)$$

$$A^{(03)} = \begin{pmatrix} C_{16} + C_{61} & C_{12} + C_{66} & C_{14} + C_{65} \\ C_{66} + C_{21} & C_{62} + C_{26} & C_{64} + C_{25} \\ C_{56} + C_{41} & C_{52} + C_{46} & C_{54} + C_{45} \end{pmatrix}, \qquad (A.16)$$

$$\mathbf{A}^{(04)} = \begin{pmatrix} -\varrho\partial^2/\partial t^2 & 0 & 0 \\ 0 & -\varrho\partial^2/\partial t^2 & 0 \\ 0 & 0 & -\varrho\partial^2/\partial t^2 \end{pmatrix}, \tag{A.17}$$

$$\mathbf{A}^{(11)} = \begin{pmatrix} C_{15}+C_{51} & C_{14}+C_{56} & C_{13}+C_{55} \\ C_{65}+C_{41} & C_{64}+C_{46} & C_{63}+C_{45} \\ C_{55}+C_{31} & C_{54}+C_{36} & C_{53}+C_{35} \end{pmatrix}, \tag{A.18}$$

$$\mathbf{A}^{(12)} = \begin{pmatrix} C_{65}+C_{56} & C_{64}+C_{52} & C_{63}+C_{54} \\ C_{25}+C_{46} & C_{24}+C_{42} & C_{23}+C_{44} \\ C_{45}+C_{36} & C_{44}+C_{32} & C_{43}+C_{34} \end{pmatrix}, \tag{A.19}$$

$$\mathbf{A}^{(2)} = \begin{pmatrix} C_{55} & C_{54} & C_{53} \\ C_{45} & C_{44} & C_{43} \\ C_{35} & C_{34} & C_{33} \end{pmatrix}. \tag{A.20}$$

A.6 Matrix for Mindlin Laminated Plate Theory

The equation of motion within the Mindlin Laminated Plate Theory for the displacement vector can be given in an explicit form (2.29), with matrix $\mathbf{T}_M$ defined as

$$
\begin{aligned}
T_{M,11} &= A_{11}\mathrm{d}_{xx} + 2A_{16}\mathrm{d}_{xy} + A_{66}\mathrm{d}_{yy} - I_0\mathrm{d}_{tt}, \\
T_{M,12} &= A_{16}\mathrm{d}_{xx} + (A_{12} + A_{66})\mathrm{d}_{xy} + A_{26}\mathrm{d}_{yy}, \\
T_{M,13} &= 0, \quad T_{M,14} = B_{11}\mathrm{d}_{xx} + 2B_{16}\mathrm{d}_{xy} + B_{66}\mathrm{d}_{yy} - I_1\mathrm{d}_{tt}, \\
T_{M,15} &= B_{16}\mathrm{d}_{xx} + (B_{12} + B_{66})\mathrm{d}_{xy} + B_{26}\mathrm{d}_{yy}, \quad T_{M,21} = T_{M,12}, \\
T_{M,22} &= A_{66}\mathrm{d}_{xx} + 2A_{26}\mathrm{d}_{xy} + A_{22}\mathrm{d}_{yy} - I_0\mathrm{d}_{tt}, \quad T_{M,23} = 0, \\
T_{M,24} &= T_{M,15}, \quad T_{M,25} = B_{66}\mathrm{d}_{xx} + 2B_{26}\mathrm{d}_{xy} + B_{22}\mathrm{d}_{yy} - I_1\mathrm{d}_{tt}, \\
T_{M,31} &= T_{M,13}, \quad T_{M,32} = T_{M,23}, \quad T_{M,33} = \kappa_1^2 A_{55}\mathrm{d}_{xx} + 2\kappa_1\kappa_2 A_{45}\mathrm{d}_{xy} + \kappa_2^2 A_{44}\mathrm{d}_{yy} - I_0\mathrm{d}_{tt}, \\
T_{M,34} &= \kappa_1^2 A_{55}\mathrm{d}_x + \kappa_1\kappa_2 A_{45}\mathrm{d}_y, \quad T_{M,35} = \kappa_1\kappa_2 A_{45}\mathrm{d}_x + \kappa_2^2 A_{44}\mathrm{d}_y, \\
T_{M,41} &= T_{M,14}, \quad T_{M,42} = T_{M,24}, \quad T_{M,43} = -T_{M,34}, \\
T_{M,44} &= D_{11}\mathrm{d}_{xx} + 2D_{16}\mathrm{d}_{xy} + D_{66}\mathrm{d}_{yy} - \kappa_1^2 A_{55} - I_2\mathrm{d}_{tt}, \\
T_{M,45} &= D_{16}\mathrm{d}_{xx} + (D_{12} + D_{66})\mathrm{d}_{xy} + D_{26}\mathrm{d}_{yy} - \kappa_1\kappa_2 A_{45}, \\
T_{M,51} &= T_{M,15}, \quad T_{M,52} = T_{M,25}, \quad T_{M,53} = -T_{M,35}, \quad T_{M,54} = T_{M,45}, \\
T_{M,55} &= D_{66}\mathrm{d}_{xx} + 2D_{26}\mathrm{d}_{xy} + D_{22}\mathrm{d}_{yy} - \kappa_2^2 A_{44} - I_2\mathrm{d}_{tt},
\end{aligned}
$$

with $\mathrm{d}_x = \partial/\partial x$, $\mathrm{d}_y = \partial/\partial y$, $\mathrm{d}_{xx} = \partial^2/\partial x^2$, $\mathrm{d}_{xy} = \partial^2/\partial x\partial y$, $\mathrm{d}_{yy} = \partial^2/\partial y^2$ and $\mathrm{d}_{tt} = \partial^2/\partial t^2$.

A.7 Matrix for Classical Laminated Plate Theory

Similarly to the MLPT, the equation of motion within the Classical Laminated Plate Theory for displacement vector can be given in an explicit form (2.36), with the matrix

$\mathbf{T}_C$ defined as

$$
\begin{aligned}
T_{C,11} &= A_{11}d_{xx} + 2A_{16}d_{xy} + A_{66}d_{yy} - I_0 d_{tt}, \\
T_{C,12} &= A_{16}d_{xx} + (A_{12} + A_{66})d_{xy} + A_{26}d_{yy}, \\
T_{C,13} &= -B_{11}d_{xxx} - 3B_{16}d_{xxy} - (B_{12} + 2B_{66})d_{xyy} - B_{26}d_{yyy}, \\
T_{C,21} &= T_{C,12}, \quad T_{C,22} = A_{66}d_{xx} + 2A_{26}d_{xy} + A_{22}d_{yy} - I_0 d_{tt}, \\
T_{C,23} &= -B_{16}d_{xxx} - (B_{12} + 2B_{66})d_{xxy} - 3B_{26}d_{xyy} - B_{22}d_{yyy}, \\
T_{C,31} &= -T_{C,13}, \quad T_{C,32} = -T_{C,23}, \\
T_{C,33} &= -D_{11}d_{xxxx} - D_{22}d_{yyyy} - 2(D_{12} + D_{66})d_{xxyy} - 4D_{16}d_{xxxy} - 4D_{26}d_{xyyy} - I_0 d_{tt},
\end{aligned}
$$

with the same notations for partial derivative operator as for a Mindlin plate, i.e. $d_{xxxy} = \partial^4/\partial x^3 \partial y$ and so on.

A.8 Wavenumber domain Green's matrix for an isotropic plate

In case of an isotropic single-layered plate Green's matrix for the elastodynamic problem (2.5), (2.10), (2.12) is described by matrix (3.65), where functions M, P, S, R and N are given as follows [10]:

$$
M = M_1/\Delta, \quad P = P_1/\Delta, \quad S = S_1/\Delta, R = R_1/\Delta, \quad N = N_1/\Delta, \tag{A.21}
$$

$$
\begin{aligned}
M_1(\alpha,z) = {}& -i\sigma_2/\alpha^2 \left\{ \alpha^2(\sigma_1\sigma_2\eta^2 \sinh\sigma_2 z + \eta^4 \sinh\sigma_1 z) \right. \\
& - \alpha^2\eta^4 \cosh\sigma_2 h \sinh\sigma_1(z+h) + \alpha^4\sigma_1\sigma_2 \sinh\sigma_2 h \cosh\sigma_1(z+h) \\
& \left. - \alpha^2\sigma_1\sigma_2\eta^2 \cosh\sigma_1 h \sinh\sigma_2(z+h) + \eta^6 \sinh\sigma_1 h \cosh\sigma_2(z+h) \right\},
\end{aligned} \tag{A.22}
$$

$$
N_1(\alpha,z) = i\cosh\sigma_2(z+h)/(G\alpha^2\sigma_2 \sinh\sigma_2 h), \tag{A.23}
$$

$$
\begin{aligned}
P_1(\alpha,z) = {}& -\sigma_1\sigma_2[\eta^2\alpha^2 \cosh\sigma_1 z + \eta^4 \cosh\sigma_2 z] \\
& - \alpha^2\sigma_1^2\sigma_2^2 \sinh\sigma_1 h \sinh\sigma_2(z+h) + \sigma_1\sigma_2\eta^4 \cosh\sigma_1 h \cosh\sigma_2(z+h) \\
& + \alpha^2\sigma_1\sigma_2\eta^2 \cosh\sigma_2 h \cosh\sigma_1(z+h) - \eta^6 \sinh\sigma_2 h \sinh\sigma_1(z+h),
\end{aligned} \tag{A.24}
$$

$$
\begin{aligned}
S_1(\alpha,z) = {}& -i \left\{ \sigma_1\sigma_2(\alpha^2\eta^2 \cosh\sigma_2 z + \eta^4 \cosh\sigma_1 z) \right. \\
& - \sigma_1\sigma_2\eta^4 \cosh\sigma_2 h \cosh\sigma_1(z+h) + \alpha^2\sigma_1^2\sigma_2^2 \sinh\sigma_2 h \sinh\sigma_1(z+h) \\
& \left. - \alpha^2\sigma_1\sigma_2\eta^2 \cosh\sigma_1 h \cosh\sigma_2(z+h) + \eta^6 \sinh\sigma_1 h \sinh\sigma_2(z+h) \right\},
\end{aligned} \tag{A.25}
$$

$$
\begin{aligned}
R_1(\alpha,z) = {}& \sigma_1 \left\{ -\alpha^2(\sigma_1\sigma_2\eta^2 \sinh\sigma_1 z + \eta^4 \sinh\sigma_2 z) \right. \\
& + \alpha^2\sigma_1\sigma_2\eta^2 \cosh\sigma_2 h \sinh\sigma_1(z+h) - \eta^6 \sinh\sigma_2 h \cosh\sigma_1(z+h) \\
& \left. + \alpha^2\eta^4 \cosh\sigma_1 h \sinh\sigma_2(z+h) - \alpha^4\sigma_1\sigma_2 \sinh\sigma_1 h \cosh\sigma_2(z+h) \right\},
\end{aligned} \tag{A.26}
$$

$$\Delta(\alpha) \;=\; 2G\left[-2\alpha^2\sigma_1\sigma_2\eta^4 - (\eta^8 + \alpha^4\sigma_1^2\sigma_2^2)\sinh\sigma_1 h \sinh\sigma_2 h \right. \tag{A.27}$$

$$\left. + \; 2\alpha^2\sigma_1\sigma_2\eta^4 \cosh\sigma_1 h \cosh\sigma_2 h \right\}, \quad \eta^2 = \alpha^2 - 0.5\zeta_2^2,$$

where $\zeta_2^2 = \omega^2/c_s^2$, G is the shear modulus of the plate (see Equation (A.13)), $\sigma_1^2 = \alpha^2 - \omega^2/c_l^2$, $\sigma_2^2 = \alpha^2 - \omega^2/c_s^2$ and $c_l = \sqrt{C_{11}/\varrho}$, $c_s = \sqrt{G/\varrho}$. Note that in a general case of N-layered isotropic laminated structure the functions M, P, S, R, N and Δ can be evaluated numerically only.

A.9 Properties of composite materials under study

Table A.1: Materials properties (elastic constants in 10^{11} Pa, density in 10^3 kg/m^3)

	C_{11}	$C_{12} = C_{13}$	$C_{22} = C_{33}$	C_{23}	C_{44}	$C_{55} = C_{66}$	ϱ
AS4/3502 [149]	1.308	0.053	0.13	0.046	0.038	0.06	1.578
CFRP-T700GC/M21 [20]	1.234	0.055	0.115	0.064	0.026	0.045	1.6
IM7-Cycom-977-3 [120]	1.528	0.11	0.232	0.176	0.028	0.033	1.558
Graphite-epoxy I [26]	1.607	0.064	0.139	0.064	0.035	0.07	1.6
Graphite-epoxy II [109]	1.607	0.064	0.139	0.069	0.035	0.07	1.578

Table A.2: Cut-off frequencies (in KHz $\cdot$ mm) of lowest higher-order Lamb wave modes in different laminated plates

Stacking sequence	Material	A_1, SH_1	MLPT, $\kappa_1 = \kappa_2 = \sqrt{5/6}$	S_1
$[45_6/-45_6]_s$	AS4/3502	797, 922	883	1434
$[45/-45/0/90]_s$	AS4/3502	835, 872	883	1434
$[0/90]_s$	CFRP-T700GC/M21	662, 785.5	750	1343
$[0/90/0/90]$	CFRP-T700GC/M21	716, 1343	750	1423
$[0]$	Graphite-epoxy I	739, 1051	745, 1058	1475
$[0/90/45/-45]_s$	Graphite-epoxy II	833, 887	921	1488
$[0/45/-45/90]_{2s}$	IM7-Cycom-977-3	689, 700	701	1389

B Mathematical background

B.1 Residue theorem and evaluation of residues on PC

Numerical evaluation of residues:

$$\operatorname{res} f(\alpha)\Big|_{\alpha=k} = \left(f(k+\delta) - f(k-\delta)\right)\delta/2 + \mathcal{O}(\delta^2). \tag{B.1}$$

Note that δ should be of order $\delta \sim 10^3\varepsilon$ if ε is the precision of the finding of a pole [10].

The residues of even and odd functions are odd and even functions, respectively:

$$\operatorname{res} f(\alpha)\Big|_{\alpha=-k} = -\operatorname{res} f(\alpha)\Big|_{\alpha=k}, \quad \text{if } f(-\alpha) = f(\alpha), \tag{B.2}$$

$$\operatorname{res} f(\alpha)\Big|_{\alpha=-k} = \operatorname{res} f(\alpha)\Big|_{\alpha=k}, \quad \text{if } f(-\alpha) = -f(\alpha).$$

B.2 Asymptotic methods

Below the stationary phase method is summarized, which is also known as the method of steepest descent [28]. Consider the integral

$$F(r) = \int_a^b f(x)\,e^{rS(x)}\,dx, \tag{B.3}$$

for $r \to \infty$, where functions $f(x)$, $S(x)$ are smooth functions for $x \to x_0$, $x \in [a,b]$. If $\max\limits_{x_0\in(a,b)} \operatorname{Re} S(x)$ is reached in an internal point x_0, where $S'(x_0) = 0$, $S''(x_0) \neq 0$, then for $r \to \infty$, the representation

$$F(r) \equiv \int_a^b f(x)\,e^{rS(x)}\,dx \sim e^{rS(x_0)} \sum_{k=0}^{\infty} c_k r^{-k-1/2} \tag{B.4}$$

is valid, which is differentiable with respect to r any times. Here,

$$c_k = \frac{\Gamma\!\left(k+\dfrac{1}{2}\right)}{(2k)!} \left(\frac{d}{dx}\right)^k \left[f(x) \left(\frac{2(S(x_0) - S(x))}{(x - x_0)^2}\right)^{-k-1/2} \right] \tag{B.5}$$

or in other form

$$F(r) \sim e^{rS(x_0)} \sum_{k=0}^{\infty} \frac{b_{2k}(r)}{(2k)!} \left[-\frac{rS''(x_0)}{2} \right]^{-k-1/2} \Gamma(k + \frac{1}{2}), \tag{B.6}$$

where

$$b_k = \left(\frac{\mathrm{d}}{\mathrm{d}x} \right)^k \left(f(x) \exp \left[r(S(x) - S(x_0) - \frac{(x - x_0)^2}{2} S''(x_0)) \right] \right) \Bigg|_{x=x_0}. \tag{B.7}$$

The main term in (B.4) is given by

$$F(r) \sim e^{rS(x_0)} f(x_0) \sqrt{-\frac{2\pi}{-rS''(x_0)}} + \mathcal{O}(r^{-3/2}). \tag{B.8}$$

The contribution of an endpoint $x = a$, if $\max_{x \in [a,b]} S(x)$ is reached in $x = a$ and for $x \to a$ the functions $f(x)$ and $S(x)$ are smooth, and $S'(a) \neq 0$, then for $r \to \infty$, the relation

$$F(r) \sim e^{rS(a)} \sum_{k=0}^{\infty} c_k r^{-k-1} \tag{B.9}$$

holds, where

$$c_k = -\left(-\frac{1}{S'(x)} \frac{\mathrm{d}}{\mathrm{d}x} \right)^k \left(\frac{f(x)}{S'(x)} \right) \Bigg|_{x=a}. \tag{B.10}$$

The main term in (B.9) is given by

$$F(r) \sim e^{rS(a)} r^{-1} \left(\frac{f(x)}{S'(x)} \right) \Bigg|_{x=a} + \mathcal{O}(r^{-2}). \tag{B.11}$$

B.3 Properties of Bessel, Hankel and Struve functions

B.3.1 Bessel and Hankel functions

Jacobi-Anger expansion is a series representation of exponentials of trigonometric functions

$$e^{i\alpha \cos \theta} = \sum_{n=-\infty}^{\infty} i^n J_n(\alpha) e^{in\theta} = J_0(\alpha) + 2 \sum_{n=1}^{\infty} i^n J_n(\alpha) \cos(n\theta), \tag{B.12}$$

$$e^{i\alpha \sin \theta} = \sum_{n=-\infty}^{\infty} J_n(\alpha) e^{in\theta},$$

where $J_n(\alpha)$ are Bessel functions. The Bessel functions exhibit the symmetry properties

$$J_n(-\alpha) = (-1)^n J_n(\alpha), \tag{B.13}$$

and the asymptotic properties

$$J_n(\alpha) = \sqrt{\frac{2}{\pi\alpha}}\cos\left(\alpha - \frac{n\pi}{2} - \frac{\pi}{4}\right) + \mathcal{O}(r^{-3/2}), \quad |\alpha| \gg 1 \tag{B.14}$$

for $\alpha \to \infty$.

The Bessel function can be separated into the sum of Hankel functions $H_n^{(j)}(\alpha)$, $j = 1, 2$:

$$J_n(\alpha) = \frac{H_n^{(1)}(\alpha) + H_n^{(2)}(\alpha)}{2}. \tag{B.15}$$

Hankel functions of the second kind are connected with Hankel function of the first kind by

$$H_n^{(2)}(-z) = (-1)^{n+1} H_n^{(1)}(z). \tag{B.16}$$

The asymptotic forms of Hankel functions are given as

$$H_m^{(1)}(z) = (-\mathrm{i})^m \sqrt{\frac{2}{\pi z}}\, \mathrm{e}^{\mathrm{i}(z-\pi/4)} + \mathrm{O}(r^{-3/2}), \tag{B.17}$$

$$H_m^{(2)}(z) = \mathrm{i}^m \sqrt{\frac{2}{\pi z}}\, \mathrm{e}^{-\mathrm{i}(z-\pi/4)} + \mathrm{O}(r^{-3/2}),$$

B.3.2 Struve functions and some typical integrals

The integrals (5.41) derived in section 5.4 for computation of out-of-plane displacement for isotropic laminate can be calculated using the Jacobi-Anger expansion (B.12):

$$\int_{\varphi+\pi/2}^{\varphi+3\pi/2} \mathrm{e}^{-\mathrm{i}\alpha r \cos(\gamma-\varphi)}d\gamma = \pi\left(J_0(\alpha r) + \mathrm{i}\mathbf{H}_0(\alpha r)\right), \tag{B.18}$$

$$\int_{\varphi-\pi/2}^{\varphi+\pi/2} \mathrm{e}^{-\mathrm{i}\alpha r \cos(\gamma-\varphi)}d\gamma = \pi\left(J_0(\alpha r) - \mathrm{i}\mathbf{H}_0(\alpha r)\right),$$

where $\mathbf{H}_0(\alpha r)$ is the so-called Struve function of zero order, defined as [105]

$$\mathbf{H}_0(\alpha) = \frac{4}{\pi}\sum_{n=0}^{\infty}\frac{J_{2n+1}(\alpha)}{2n+1}. \tag{B.19}$$

Similarly, the integrals for computation of in-plane radial displacement for isotropic laminate can be brought to an analytical representation while using the Jacobi-Anger expansion (B.12):

$$\int_{\varphi+\pi/2}^{\varphi+3\pi/2} \cos(\gamma-\varphi)\mathrm{e}^{-\mathrm{i}\alpha r \cos(\gamma-\varphi)}d\gamma = \pi\left(J_1(\alpha r) + \mathrm{i}\left(\mathbf{H}_1(\alpha r) - \frac{2}{\pi}\right)\right), \tag{B.20}$$

$$\int_{\varphi-\pi/2}^{\varphi+\pi/2} \cos(\gamma-\varphi)\mathrm{e}^{-\mathrm{i}\alpha r \cos(\gamma-\varphi)}d\gamma = \pi\left(J_1(\alpha r) - \mathrm{i}\left(\mathbf{H}_1(\alpha r) - \frac{2}{\pi}\right)\right),$$

where $\mathbf{H}_1(\alpha r)$ is a Struve function of first order and it is represented by [105]

$$\mathbf{H}_1(\alpha) = \frac{2}{\pi}(1 - J_0(\alpha)) + \frac{4}{\pi}\sum_{n=1}^{\infty}\frac{J_{2n}(\alpha)}{4n^2 - 1}.$$
(B.21)

The Struve functions of an odd order are even functions:

$$\mathbf{H}_n(-\alpha) = (-1)^{n+1}\mathbf{H}_n(\alpha).$$
(B.22)

The analytical expressions obtained for the integrals (B.18), (B.20) have the asymptotic properties

$$\pi\left(J_1(\alpha r) + (-1)^{(j+1)}\mathrm{i}(\mathbf{H}_1(\alpha r) - \tfrac{2}{\pi})\right) = \pi H_1^{(j)}(\alpha r)$$
$$+2\mathrm{i}(-1)^{(j+1)}\left(\frac{1}{\alpha^2 r^2} - \frac{3}{\alpha^4 r^4} + \mathcal{O}(r^{-6})\right),$$
(B.23)
$$\pi\left(J_0(\alpha r) + (-1)^{(j+1)}\mathrm{i}\mathbf{H}_0(\alpha r)\right) = \pi H_0^{(j)}(\alpha r)$$
$$+2\mathrm{i}(-1)^{(j+1)}\left(\frac{1}{\alpha r} - \frac{1}{\alpha^3 r^3} + \mathcal{O}(r^{-5})\right),$$

where $j = 1$ for the first and $j = 2$ for the second integrals in Equations (B.18), (B.20), respectively. Note that the last asymptotics can be directly derived by applying to integrals (B.18), (B.20) the method of steepest descent (Appendix B.2).

List of Figures

List of Tables

Bibliography

[1] *D01AKF subroutine. NAG Library Manual. The Numerical Algorithms Group Ltd,*
Oxford, UK. 2009.

[2] J. Achenbach. *Wave Propagation in Elastic Solids.* North-Holland Publishing Company, Amsterdam, 1973.

[3] L. Ahlfors. *Complex analysis: an introduction to the theory of analytic functions of one complex variable.* McGraw-Hill, 1979.

[4] D. Alleyne. *The Nondestructive Testing of Plates Using Ultrasonic Lamb Waves.* Dissertation, Imperial College of Science, Technology and Medicine London, 1991.

[5] D. Alleyne and P. Cawley. *Two dimensional Fourier transform method for the measurement of propagating mulitmode signals.* Journal of the Acoustical Society of America, 89:1159–1168, 1991.

[6] D. Alleyne, M. Lowe and P. Cawley. *Inspection of chemical plant pipework using Lamb waves: defect sensitivity and field experience.* In Review of Progress in Quantitative NDE, 15:1859–1866, 1996.

[7] B. Auld. *Acoustic Fields and Waves in Solids.* Wiley-Interscience, New York, 1973.

[8] B. Auld. *Acoustic fields and waves in solids: Volumes I & II.* R.E. Kreiger Publishing Co., Florida, 2nd Ed., 1990.

[9] V. Babeshko and P. Syromyatnikov. *A method for the construction of the Fourier symbol of the Green matrix for multi-layered electroelastic half-space.* Mechanics of Solids, 37 (5):27–37, 2002.

[10] V. Babeshko, E. Glushkov and Z. Zinchenko. *Dynamics of Inhomogeneous Linearly Elastic Media [In Russian].* Nauka, Moscow, 1989.

[11] V. Babeshko, S. Ratner and P. Syromyatnikov. *On Mixed Problems for Thermo-electroelastic Media with Discontinuous Boundary Conditions [In Russian].* Doklady Physics, 52:90–95, 2007. doi: 10.1134/S102833580702005X.

[12] H. Bai, J. Zhu, A. Shah and N. Popplewell. *Three-dimensional steadystate Green function for a layered isotropic plate.* Journal of Sound and Vibration, 269:251–271, 2004. doi: 10.1016/S0022-460X(03)00071-3.

[13] S. Banerjee, W. Prosser and A. Mal. *Calculation of the Response of a Composite Plate to Localized Dynamic Surface Loads Using a New Wave Number Integral Method.* Journal of Applied Mechanics, 72:18–24, 2005. doi: 10.1115/1.1828064.

[14] J. Bao. *Lamb Wave Generation and Detection with Piezoelectric Wafer Active Sebsors.* Dissertation, University of South Carolina, Columbia, 2003.

[15] J.-M. Berthelot. *Composite Materials, Mechanical Behavior and Structural Analysis.* Springer New York, 1999.

[16] J.-M. Berthelot. *Mechanical Behaviour of Composite Materials and Structures.* IS-MANS, Le Mans, France, 2007.

[17] S. Bhalla and A. Gupta. *Piezoelectric Ceramics. Chapter: Modelling Shear Lag Phenomenon for Adhesively Bonded Piezo-Impedance Transducers.* InTech, 2010.

[18] D. Bray and D. McBride. *Nondestructive testing techniques.* New York, Wiley, 1992.

[19] E. Carrera. *Theories and Finite Elements for Multilayered, Anisotropic, Composite Plates and Shells.* Archives of Computational Methods in Engineering, 9 (2):87–140, 2002.

[20] B. Chapuis, N. Terrien and D. Royer. *Excitation and focusing of Lamb waves in a multilayered anisotropic plate.* Journal of Acoustical Society of America, 127 (1): 198–203, 2010. doi: 10.1121/1.3263607.

[21] D. Chimenti. *Guided waves in plates and their use in materials characterization.* Applied Mechanics Reviews, 50:247–284, 1997.

[22] D. Chimenti and R. Martin. *Nondestructive evaluation of composite laminates by leaky Lamb waves.* Ultrasonics, 29:13–21, 1991.

[23] Y. Cho and J. Rose. *A boundary element solution for a mode conversion study on the edge reflection of Lamb waves.* Journal of the Acoustical Society of America, 99: 2097–2109, 1996.

[24] E. Christoffel. *A boundary element solution for a mode conversion study on the edge reflection of Lamb waves.* Annali di Matematica Pura ed Applicata, 8:193, 1877.

[25] E. Crawley and J. de Luis. *Use of Piezoelectric Actuators as Elements of Intelligent Structures.* AIAA Journal, 25 (10):1373–1385, 1987. doi: 10.2514/3.9792.

[26] S. Datta and A. Shah. *Elastic Waves in Composite Media and Structures.* CRC Press, 2009.

[27] L. Duquenne, E. Moulin, J. Assaad and S. Grondel. *Transient modeling of Lamb waves generated in viscoelastic materials by surface bonded piezoelectric transducers.* Journal of the Acoustical Society of America, 116 (1):133–141, 2004.

[28] M. Fedoryuk. *The Saddle-Point Method [In Russian]*. Nauka, Moscow, 1977.

[29] M. Fedoryuk. *Asymptotics, Integrals and Series [In Russian]*. Nauka, Moscow, 1987.

[30] C.-P. Fritzen and P. Kraemer. *Self-diagnosis of smart structures based on dynamical properties*. Mechanical Systems and Signal Processing, 23:1830–1845, 2009. doi: 10.1016/j.ymssp.2009.01.006.

[31] J. Galan and R. Abascal. *Numerical simulation of Lamb wave scattering in semi-infinite plates*. International Journal for Numerical Methods in Engineering, 53: 1145–1173, 2002.

[32] V. Giurgiutiu. *Tuned Lamb Wave Excitation and Detection with Piezoelectric Wafer Active Sensors for Structural Health Monitoring*. Journal of Intelligent Material Systems and Structures, 16:291–305, 2005. doi: 10.1177/1045389X05050106.

[33] V. Giurgiutiu. *Structural Health Monitoring with Piezoelectric Wafer Active Sensors*. Elsevier Academic Press, 2008.

[34] E. Glushkov and N. Glushkova. *On the efficient implementation of the integral equation method in elastodynamics*. Journal of Computational Acoustics, 9:889–897, 2001. doi: 10.1142/S0218396X01001169.

[35] E. Glushkov and E. Kirillova. *A dynamic mixed problem for a packet of elastic layers*. Journal of Applied Mathematics and Mechanics, 62 (3):419–425, 1998.

[36] E. Glushkov, N. Glushkova, A. Ekhlakov and E. Shapar. *An analytically based computer model for surface measurements in ultrasonic crack detection*. Wave Motion, 43 (6):458–473, 2006. doi: 10.1016/j.wavemoti.2006.03.002.

[37] E. Glushkov, N. Glushkova, O. Kvasha and W. Seemann. *Integral equation based modeling of the interaction between piezoelectric patch actuators and an elastic substrate*. Smart Materials and Structures, 16:650–664, 2007. doi: 10.1088/0964-1726/16/3/012.

[38] E. Glushkov, N. Glushkova, O. Kvasha, D. Kern and W. Seemann. *Guided Wave Generation and Sensing in an Elastic Beam Using MFC Piezoelectric Elements: Theory and Experiment*. Journal of Intelligent Material Systems and Structures, 21:1617–1625, 2010. doi: 10.1177/1045389X10385486.

[39] E. Glushkov, N. Glushkova, O. Kvasha and R. Lammering. *Selective Lamb mode excitation by piezoelectric coaxial ring actuators*. Smart Materials and Structures, 19: 035018, 2010. doi: 10.1088/0964-1726/19/3/035018.

[40] E. Glushkov, N. Glushkova and A. Eremin. *Forced wave propagation and energy distribution in anisotropic laminate composites*. Journal of Acoustical Society of America, 129 (5):2923–2934, 2011. doi: 10.1121/1.3559699.

[41] E. Glushkov, N. Glushkova, R. Lammering, A. Eremin and M. Neumann. *Lamb wave excitation and propagation in elastic plates with surface obstacles: proper choice of central frequencies.* Smart Materials and Structures, 20:015020, 2011. doi: 10.1088/0964-1726/20/1/015020.

[42] E. Glushkov, N. Glushkova, W. Seemann and O. Kvasha. *The excitation of waves by piezoelectric patch actuators arranged symmetrically on both surfaces of an elastic layer.* Journal of Applied Mathematics and Mechanics, 75 (1):56–64, 2011. doi: 10.1016/j.jappmathmech.2011.04.008.

[43] Y. Glushkov, N. Glushkova and A. Krivonos. *The excitation and propagation of elastic waves in multilayered anisotropic composites.* Journal of Applied Mathematics and Mechanics, 74 (3):419–432, 2010. doi: 10.1016/j.jappmathmech.2010.07.005.

[44] S. Gopalakrishnan, A. Chakraborty and D. R. Mahapatra. *Spectral Finite Element Method. Wave Propagation, Diagnostics and Control in Anisotropic and Inhomogeneous Structures.* Springer, 2007. doi: 10.1007/978-1-84628-356-7.

[45] K. Graff. *Elastic Waves in Solids.* Oxford University Press, New York, 1973.

[46] N. Haskell. *Dispersion of surface waves on multilayered media.* Bulletin of the Seismological Society of America, 43:17–34, 1953.

[47] T. Hay, L. Wei, J. Rose and T. Hayashi. *Rapid Inspection of Composite Skin-Honeycomb Core Structures with Ultrasonic Guided Waves.* Journal of Composite Materials, 37:929–939, 2003. doi: 10.1177/0021998303037010005.

[48] T. Hayashi. *Guided Wave Dispersion Curves Derived with a Semianalytical Finite Element Method and Its Applications to Nondestructive Inspection.* Japanese Journal of Applied Physics, 47 (5):3865–3870, 2008. doi: 10.1143/JJAP.47.3865.

[49] T. Hayashi and K. Kawashima. *Multiple reflections of Lamb waves at a delamination.* Ultrasonics, 40:193–197, 2002.

[50] T. Hayashi and J. L. Rose. *Guided wave simulation and visualization by a Semi-Analytical Finite Element Method.* Materials Evaluation, 61 (1):75–79, 2003.

[51] H. Hibbitt, B. Karlsson and P. Sorensen. *ABAQUS Analysis User'sManual Version 6.10. Dassault Systemes Simulia Corp.: Providence, RI, USA,* 2011.

[52] B. Hosten and M. Castaings. *Transfer matrix of multilayered absorbing and anisotropic media. Measurements and simulations of ultrasonic wave propagation through composite materials.* Journal of Acoustical Society of America, 94:1488–1495, 1993. doi: 10.1121/1.408152.

[53] H. Huang, T. Pamphile and M. Derriso. *The effect of actuator bending on Lamb wave displacement fields generated by a piezoelectric patch.* Smart Materials and Structures, 17:055012, 2008. doi: 10.1088/0964-1726/17/5/055012.

[54] H. Huang, F. Song and X. Wang. *Quantitative Modeling of Coupled Piezo-Elastodynamic Behavior of Piezoelectric Actuators Bonded to an Elastic Medium for Structural Health Monitoring: A Review.* Sensors, 10:3681–3702, 2010. doi: 10.3390/s100403681.

[55] D. Jansena, D. Hutchinsa and J. Mottram. *Lamb wave tomography of advanced composite laminates containing damage.* Ultrasonics, 32 (2):83–90, 1994. doi: 10.1016/0041-624X(94)90015-9.

[56] R. Jones. *Mechanics of Composite Materials.* Taylor & Francis, 2nd Ed., 1999.

[57] A. Karmazin, E. Kirillova and P. Syromyatnikov. *Phase speeds of Lamb waves in multilayered anisotropic composites [In Russian].* Ecological Bulletin of Research Centers of the BSEC Member States, 2:22–31, 2009.

[58] A. Karmazin, E. Kirillova, W. Seemann and P. Syromyatnikov. *Modelling of 3D Steady-State Oscillations of Anisotropic Multilayered Structures Applying the Green's Functions.* Advances in Theoretical and Applied Mechanics, 3 (9):425–450, 2010.

[59] A. Karmazin, E. Kirillova, W. Seemann and P. Syromyatnikov. *Analysis of spatial steady-state vibrations of a layered anisotropic plate using the Green's functions. ESDA2010-25430. In Proceedings of the ASME 2010 10th Biennial Conference on Engineering Systems & Design and Analysis (ESDA 2010) July 12-14, 2010, Istanbul, Turkey,* 2010. doi: 10.1115/ESDA2010-25430.

[60] A. Karmazin, E. Kirillova, W. Seemann and P. Syromyatnikov. *Investigation of Dispersion Characteristics of Composites.* Proceedings in Applied Mathematics and Mechanics, 10 (1):503–504, 2010. doi: 10.1002/pamm.201010244.

[61] A. Karmazin, E. Kirillova, W. Seemann and P. Syromyatnikov. *Investigation of Lamb elastic waves in anisotropic multilayered composites applying the Green's matrix.* Ultrasonics, 51:17–28, 2011. doi: 10.1016/j.ultras.2010.05.003.

[62] A. Karmazin, E. Kirillova, W. Seemann and P. Syromyatnikov. *A study of time harmonic guided Lamb waves and their caustics in composite plates.* Ultrasonics, 53: 283–293, 2013. doi: 10.1016/j.ultras.2012.06.012.

[63] E. Kausel. *Wave propagation in anisotropic layered media.* International Journal for Numerical Methods in Engineering, 23:1567–1578, 1986. doi: 10.1002/nme.1620230811.

[64] E. Kausel and J. Roesset. *Frequency domain analysis of undamped systems.* ASCE Journal of Engineering Mechanics, 118:721–734, 1992.

[65] L. Knopoff. *A matrix method for elastic wave problems.* Seismological Society of America, 54 (1):431–438, 1964.

[66] Y. Kravtsov. *Complex rays and complex caustics [in Russian].* Izvestiya vuzov: Radiofizika, 10:1283, 1967.

[67] Y. Kravtsov and Y. Orlov. *Geometrical Optics of Inhomogeneous Media.* Springer-Verlag Berlin, 1990.

[68] Y. Kravtsov and Y. Orlov. *Caustics, Catastrophes and Wave Fields.* Springer New York, 1993.

[69] A. Krivonos. *Excitation and elastic wave propagation in multi-layered anisotropic composites [In Russian].* Dissertation, Kuban State University, 2010.

[70] P. Kudela, W. Ostachowicz and A. Zak. *Damage detection in composite plates with embedded PZT transducers.* Mechanical Systems and Signal Processing, 22:1327–1335, 2008.

[71] H. Lamb. *On waves in an elastic plate.* Proceedings of the Royal Society of London Series A, 93:293–312, 1917.

[72] B. Lee and W. Staszewski. *Modelling of Lamb waves for damage detection in metallic structures: Part I. Wave propagation. Part II. Wave interactions with damage.* Smart Materials and Structures, 12:804–824, 2003. doi: 10.1088/0964-1726/12/5/018.

[73] B. Lee and W. Staszewski. *Lamb wave propagation modelling for damage detection: I. Two-dimensional analysis.* Smart Materials and Structures, 16:249–259, 2007. doi: 10.1088/0964-1726/16/2/003.

[74] S. Lekhnitskii. *Theory of Elasticity of an Anisotropic Body.* Holden-Day, San Francisco, 1963.

[75] B. Lempiere. *Ultrasound and Elastic Waves. Frequently Asked Questions.* Academic Press, 2002.

[76] S. Lih and A. Mal. *On the accuracy of approximate plate theories for wave field calculations in composite laminates.* Wave Motion, 21:17–34, 1995. doi: 10.1016/0165-2125(94)00038-7.

[77] X. Lin and F. Yuan. *Diagnostic Lamb waves in an integrated piezoelectric sensor/actuator plate: analytical and experimental studies.* Smart Materials and Structures, 10:907–913, 2001. doi: 10.1088/0964-1726/10/5/307.

[78] G. Liu and J. Achenbach. *A Strip Element Method for Stress Analysis of Anisotropic Linearly Elastic Solids.* Journal of Applied Mechanics, 61:270–277, 1994.

[79] G. Liu and Z. Xi. *Elastic Waves in Anisotropic Laminates.* CRC Press, 2002.

[80] G. Liu, J. Tani, K. Watanabe and T. Ohyoshi. *Lamb Wave Propagation in Anisotropic Laminates*. Journal of Applied Mechanics, 57:923–929, 1990.

[81] G. Liu, J. Tani, T. Ohyoshi and K. Watanabe. *Characteristic of surface wave propagation along the edge of an anisotropic laminated semi-infinite plate*. Wave Motion, 13: 243–251, 1991. doi: 10.1016/0165-2125(91)90061-R.

[82] M. Lowe. *Matrix Techniques for Modeling Ultrasonic Waves in Multilayered Media*. IEEE Transactions on Ultrasonics, Ferroelectrics, and Frequency Control, 42 (4): 525–542, 1995. doi: 10.1109/58.393096.

[83] J. Mackerle. *Finite-element modelling of non-destructive material evaluation: a bibliography*. Modelling and Simulation in Materials Science and Engineering, 7: 107–145, 1999.

[84] A. Mal. *Wave propagation in layered composite laminates under periodic surface loads*. Wave Motion, 10:257–266, 1988. doi: 10.1016/0165-2125(88)90022-4.

[85] H. Maris. *Enhancement of heat pulses in crystals due to elastic anisotropy*. Journal of Acoustical Society of America, 50:812–818, 1971.

[86] P. Marks. *Aviation – The shape of wings to come*. New Scientist, 2005.

[87] H. Matt. *Structural Diagnostics of CFRP Composite Aircraft Components by Ultrasonic Guided Waves and Built-In Piezoelectric Transducers*. Dissertation, University of California, Columbia, 2006.

[88] A. Maznev, A. Lomonosov, P. Hess and A. Kolomenskii. *Anisotropic effects in surface acoustic wave propagation from a point source in a crystal*. The European Physical Journal B, 35:429–439, 2003. doi: 10.1140/epjb/e2003-00295-y.

[89] J. Miklowitz. *The Theory of Elastic Waves and Waveguides*. North-Holland Publishing Company, Amsterdam, 1978.

[90] R. Mindlin. *Waves and vibrations in isotropic elastic plates*. Pergamon Press, New York, 1960.

[91] R. Mindlin. *An Introduction to the Mathematical Theory of Vibrations of Elastic Plates*. World Scientific, 2006.

[92] E. Moulin, J. Assaad and C. Delebarre. *Modeling of Lamb waves generated by integrated transducers in composite plates using a coupled finite element-normal modes expansion method*. Journal of the Acoustical Society of America, 107:87–94, 2000.

[93] D. Muller. *A Method for Solving Algebraic Equations Using an Automatic Computer*. Mathematical Tables and Other Aids to Computation, 10 (56):208–215, 1956.

[94] A. Nayfeh. *The general problem of elastic wave propagation in multilayered anisotropic media.* Journal of Acoustical Society of America, 89:1521–1531, 1991. doi: 10.1121/1.400988.

[95] A. Nayfeh. *Wave Propagation in Layered Anisotropic Media with Applications to Composites.* Elsevier Science, 1995.

[96] G. Neau. *Lamb waves in anisotropic viscoelastic plates. Study of the wave fronts and attenuation.* Dissertation, L'Universite de Bordeaux, 2003.

[97] C. Ng and M. Veidt. *A Lamb-wave-based technique for damage detection in composite laminates.* Smart Materials and Structures, 18:074006, 2009. doi: 10.1088/0964-1726/18/7/074006.

[98] C. Ng and M. Veidt. *Scattering of the fundamental anti-symmetric Lamb wave at delaminations in composite laminates.* Journal of Acoustical Society of America, 129 (3):1288–1296, 2011. doi: 10.1121/1.3533741.

[99] J. Nieuwenhuis and J. Neumann. *Generation and detection of guided waves using PZT wafer transducers.* IEEE Transactions on Ultrasonics, Ferroelectrics and Frequency Control, 52(11):2103–2111, 2005.

[100] B. Noble. *Methods Based on the Wiener-Hopf Technique for the Solution of Partial Differential Equations.* Pergamon Press, New York, 1959.

[101] M. Onoe. *Study of the branches of the velocity dispersion equations of elastic plates and rods.* Report Joint Committee on Ultrasonics of the Institute of Electrical Communication Engineers and Acoustical Society of Japan, 1955.

[102] B. Pavlakovic and M. Lowe. *Disperse software manual Version 2.0.1 6B, Imperial College, London, UK,* 2003.

[103] C. Peng and M. Toksoz. *Optimal absorbing boundary condition for finite difference modeling of acoustic and elastic wave propagation.* Journal of the Acoustical Society of America, 95:733–745, 1994.

[104] H. Peng, G. Meng and F. Li. *Modeling of wave propagation in plate structures using three-dimensional spectral element method for damage detection.* Journal of Sound and Vibration, 320:942–954, 2009. doi: 10.1016/j.jsv.2008.09.005.

[105] A. Prudnikov, Y. Brychkov and O. Marichev. *Integrals and Series. Special Functions [In Russian].* Nauka, Moscow, 1983.

[106] A. Purekar and D. Pines. *Damage Detection in Thin Composite Laminates Using Piezoelectric Phased Sensor Arrays and Guided Lamb Wave Interrogation.* Journal of Intelligent Material Systems and Structures, 21:995–1010, 2010. doi: 10.1177/1045389X10372003.

[107] A. Raghavan. *Guided-Wave Structural Health Monitoring*. Dissertation, The University of Michigan, Ann Arbor, 2007.

[108] A. Raghavan and C. Cesnik. *Finite-dimensional piezoelectric transducer modeling for guided wave based structural health monitoring*. Smart Materials and Structures, 14: 1448–1461, 2005. doi: 10.1088/0964-1726/14/6/037.

[109] A. Raghavan and C. Cesnik. *Modeling of Guided-wave Excitation by Finite-dimensional Piezoelectric Transducers in Composite Plates*. In *48th AIAA/ASME/ASCE/AHS/ASC Structures, Structural Dynamics, and Materials Conference, 23 – 26 April 2007, Honolulu, Hawaii*, 2007.

[110] A. Raghavan and C. Cesnik. *Review of guided-wave structural health monitoring*. The Shock and Vibration Digest, 39 (2):91–114, 2007. doi: 10.1177/0583102406075428.

[111] C. Ramadas, K. Balasubramaniam, M. Joshi and C. Krishnamurthy. *Interaction of the primary anti-symmetric Lamb mode (A0) with symmetric delaminations: numerical and experimental studies*. Smart Materials and Structures, 18:085011, 2009. doi: 10.1088/0964-1726/18/8/08501.

[112] L. Rayleigh. *On waves propagating along the plane of an elastic solid*. Proceedings of the London Mathematical Society, 17 (1):4–11, 1885. doi: 10.1112/plms/s1-17.1.4.

[113] J. Reddy. *Mechanics of Laminated Composite Plates and Shells. Theory and Analysis*. CRC Press, 2nd Ed., 2003.

[114] S. Rokhlin and L. Wang. *Ultrasonic waves in layered anisotropic media: characterization of multidirectional composites*. International Journal of Solids and Structures, 39 (16):4133–4149, 2002. doi: 10.1016/S0020-7683(02)00363-3.

[115] S. Rokhlin, D. Chimenti and A. Nayfeh. *Topology of the complex wave spectrum in a fluid-coupled elastic layer*. Journal of the Acoustical Society of America, 85: 1074–1080, 1989.

[116] J. Rose. *Ultrasonic Waves in Solid Media*. Cambridge University Press, Cambridge, UK, 1999.

[117] D. Royer and E. Dieulesaint. *Elastic Waves in Solids I: Free and Guided Propagation*. Springer, Heidelberg, 2000.

[118] W. Rudin. *Real and complex analysis*. McGraw Hill, 1966.

[119] K. Salas and C. Cesnik. *Guided wave excitation by a CLoVER transducer for structural health monitoring: theory and experiments*. Smart Materials and Structures, 18: 075005, 2009. doi: 10.1088/0964-1726/18/7/075005.

[120] K. Salas and C. Cesnik. *Guided wave structural health monitoring using CLoVER transducers in composite materials.* Smart Materials and Structures, 19:015014, 2010. doi: 10.1088/0964-1726/19/1/015014.

[121] K. I. Salas and C. Cesnik. *Design and Characterization of a Variable-Length Piezocomposite Transducer for Structural Health Monitoring.* Journal of Intelligent Material Systems and Structures, 21:249–360, 2010. doi: 10.1177/1045389X09343219.

[122] G. Santoni. *Fundamental Studies in the Lamb-wave Interaction between Piezoelectric Wafer Active Sensor and Host Structure During Structural Health Monitoring.* Dissertation, University of South Carolina, 2010.

[123] T. Sauter. *Computation of irregularly oscillating integrals.* Applied Numerical Mathematics, 35 (3):245–264, 2000. doi: 10.1016/S0168-9274(99)00137-3.

[124] A. Schmidt. *Experimental Investigations on Nondestructive Testing Methods for Defect Detection with Double-pulse Electronic Speckle Pattern Interferometry.* Dissertation, Helmut-Schmidt-University / University of the Federal Armed Forces Hamburg, 2009.

[125] F. Schöpfer, F. Binder, A. Wöstehoff and T. Schuster. *A Mathematical Analysis of the Strip-Element Method for the Computation of Dispersion Curves of Guided Waves in Anisotropic Layered Media.* Mathematical Problems in Engineering, 2010:17, 2010. doi: 10.1155/2010/924504.

[126] F. Schubert. *Numerical time-domain modeling of linear and nonlinear ultrasonic wave propagation using finite integration techniques – theory and applications.* Ultrasonics, 42:221–229, 2004. doi: 10.1016/j.ultras.2004.01.013.

[127] W. Seemann, A. Ekhlakov, E. Glushkov, N. Glushkova and O. Kvasha. *The modeling of piezoelectrically excited waves in beams and layered substructures.* Journal of Sound and Vibration, 301:1007–1022, 2007. doi: 10.1016/j.jsv.2006.10.037.

[128] V. Seimov. *Dynamic Contact Problems [in Russian].* Naukova Dumka, Kiev, 1976.

[129] C. Simon, C. Kaczmarek and D. Royer. *Elastic wave propagation along arbitrary direction in free orthotropic plates: application to composite materials.* In Proceedings of the 4th Congress on Acoustics, 715-718, 1997.

[130] V. Sonti, S. Kim and J. Jones. *Equivalent Forces and Wavenumber Spectra of Shaped Piezoelectric Actuators.* Journal of Sound and Vibration, 187 (1):111–131, 1995. doi: 10.1006/jsvi.1995.0505.

[131] W. Staszewski, B. Lee, L. Mallet and F. Scarpa. *Structural health monitoring using scanning laser vibrometry: I. Lamb wave sensing. II. Lamb waves for damage detection.* Smart Materials and Structures, 13:251–269, 2004. doi: 10.1088/0964-1726/13/2/002.

[132] Z. Su and L. Ye. *Lecture Notes in Applied and Computational Mechanics. Vo. 48: Identification of Damage Using Lamb Waves. From Fundamentals to Applications.* Springer, 2009. doi: 10.1007/ 978-1-84882-784-4.

[133] Z. Su, L. Ye and Y. Lu. *Guided Lamb waves for identification of damage in composite structures: A review.* Journal of Sound and Vibration, 295:753–780, 2006.

[134] E. Süli and D. Mayers. *An Introduction to Numerical Analysis.* Cambridge University Press, 2003.

[135] P. Syromyatnikov. *Energy of electroelastic waves, excitated by a harmonic surface source in a piezoelectric semiorganic media [in Russian].* Dissertation, Kuban State University, 1996.

[136] B. Tang and E. Henneke. *Long wavelength approximation for Lamb wave characterization of composite laminates.* Research in Nondestructive Evaluation, 1(1):51–64, 1989.

[137] W. Thomson. *Transmission of elastic waves through a stratified solid medium.* Journal of Applied Physics, 21:89–93, 1950.

[138] M. Torres-Arredondo, H. Jung and C.-P. Fritzen. *A Study of Attenuation and Acoustic Energy Anisotropy of Lamb Waves in Multilayered Anisotropic Media for NDT and SHM applications.* In *Proceedings of the 6th International Workshop NDT in Progress, Prague, Czech Republic,* 2011.

[139] N. Umov. *Selected works: The equations of motion of the energy in bodies, Odessa (1874) [in Russian].* State Publishers of Technical-Theoretical Literature (GITTL), Moscow-Leningrad, 1950.

[140] N. Vasudevan and A. Mal. *Response of an Elastic Plate to Localized Transient Sources.* Journal of Applied Mechanics, 52:356–362, 1985.

[141] A. Velichko and P. Wilcox. *Modeling the excitation of guided waves in generally anisotropic multilayered media.* Journal of Acoustical Society of America, 121(1): 60–69, 2007. doi: 10.1121/1.2390674.

[142] K. Verma and N. Hasebe. *Wave propagation in plates of general anisotropic media in generalized thermoelasticity.* International Journal of Engineering Science, 39: 1739–1763, 2001. doi: 10.1016/S0020-7225(01)00014-3.

[143] I. Viktorov. *Rayleigh and Lamb Waves.* Plenum Press New York, 1970.

[144] I. Vorovich and V. Babeshko. *Mixed Dynamic Problems in the Theory of Elasticity for Non-classical Domains [in Russian].* Nauka, Moscow, 1979.

[145] I. Vorovich, V. Aleksandrov and V. Babeshko. *Nonclassical Mixed Problems of the Theory of Elasticity [in Russian].* Nauka, Moscow, 1974.

[146] I. Vorovich, V. Babeshko and O. Pryakhina. *Dynamics of Massive Bodies and Resonance Phenomena in Deformable Media [In Russian].* Nauch. Mir, Moscow, 1999.

[147] C.-Y. Wang and J. Achenbach. *Lamb's problem for solids of general anisotropy.* Wave Motion, 24:227–242, 1996. doi: 10.1016/S0165-2125(96)00016-9.

[148] L. Wang and S. Rokhlin. *Stable reformulation of transfer matrix method for wave propagation in layered anisotropic media.* Ultrasonics, 39:413–424, 2001.

[149] L. Wang and F. Yuan. *Group velocity and characteristic wave curves of Lamb waves in composites: Modeling and experiments.* Composites Science and Technology, 67: 1370–1384, 2007. doi: 10.1016/j.compscitech.2006.09.023.

[150] L. Wang and F. Yuan. *Lamb wave propagation in composite laminates using a higher-order plate theory. 65310I.* In *Proc. SPIE 6531,* 2007.

[151] W. Wilkie, D. Inman, J. High and R. Williams. *Recent Developments in NASA Piezocomposite Actuator Technology.* Technical report, NASA Langley Research Center, 2004.

[152] D. Worlton. *Experimental confirmation of Lamb waves at megacycle frequencies.* Journal of Applied Physics, 32:967–971, 1961.

[153] P.-C. Xu and A. Mal. *An adaptive integration scheme for irregularly oscillatory functions.* Wave Motion, 7:235–243, 1985. doi: 10.1016/0165-2125(85)90009-5.

[154] C. Yang, L. Ye, Z. Su and M. Bannister. *Some aspects of numerical simulation for Lamb wave propagation in composite laminates.* Composite Structures, 75:267–275, 2006. doi: 10.1016/j.compstruct.2006.04.034.

[155] C. Yeum, H. Sohn and J. Ihn. *Lamb wave mode decomposition using concentric ring and circular piezoelectric transducers.* Wave Motion, 48:358–370, 2011. doi: 10.1016/j.wavemoti.2011.01.001.

[156] L. Yu, G. Bottai-Santoni and V. Giurgiutiu. *Shear lag solution for tuning ultrasonic piezoelectric wafer active sensors with applications to Lamb wave array imaging.* International Journal of Engineering Science, 48:848–861, 2010. doi: 10.1016/j.ijengsci.2010.05.007.